Water Science and Technology Library

Volume 80

The aim of the Water Science and Technology Library is to provide a forum for dissemination of the state-of-the-art of topics of current interest in the area of water science and technology. This is accomplished through publication of reference books and monographs, authored or edited. Occasionally also proceedings volumes are accepted for publication in the series.

Water Science and Technology Library encompasses a wide range of topics dealing with science as well as socio-economic aspects of water, environment, and ecology. Both the water quantity and quality issues are relevant and are embraced by Water Science and Technology Library. The emphasis may be on either the scientific content, or techniques of solution, or both. There is increasing emphasis these days on processes and Water Science and Technology Library is committed to promoting this emphasis by publishing books emphasizing scientific discussions of physical, chemical, and/or biological aspects of water resources. Likewise, current or emerging solution techniques receive high priority. Interdisciplinary coverage is encouraged. Case studies contributing to our knowledge of water science and technology are also embraced by the series. Innovative ideas and novel techniques are of particular interest.

Comments or suggestions for future volumes are welcomed.

Vijay P. Singh, Department of Biological and Agricultural Engineering & Zachry Department of Civil Engineering, Texas A&M University, USA Email: vsingh@tamu.edu

More information about this series at http://www.springer.com/series/6689

Vijay P. Singh · Shalini Yadav
Ram Narayan Yadava
Editors

Energy and Environment

Select Proceedings of ICWEES-2016

Springer

Editors
Vijay P. Singh
Department of Biological and Agricultural Engineering, and Zachry Department of Civil Engineering
Texas A&M University
College Station, TX
USA

Ram Narayan Yadava
AISECT University
Hazaribagh, Jharkhand
India

Shalini Yadav
Department of Civil Engineering
AISECT University
Bhopal, Madhya Pradesh
India

ISSN 0921-092X ISSN 1872-4663 (electronic)
Water Science and Technology Library
ISBN 978-981-10-5797-7 ISBN 978-981-10-5798-4 (eBook)
https://doi.org/10.1007/978-981-10-5798-4

Library of Congress Control Number: 2017947022

Printed on acid-free paper

This Springer imprint is published by Springer Nature
The registered company is Springer Nature Singapore Pte Ltd.
The registered company address is: 152 Beach Road, #21-01/04 Gateway East, Singapore 189721, Singapore

Preface

Fundamental to sustainable economic development, functioning of healthy ecosystems, reliable agricultural productivity, dependable power generation, maintenance of desirable environmental quality, continuing industrial growth, enjoyment of quality lifestyle, and renewal of land and air resources is water. With growing population, demands for water for agriculture and industry are skyrocketing. On the other hand, freshwater resources per capita are decreasing. There is therefore a need for effective water resources management strategies. These strategies must also consider the nexus between water, energy, environment, food, and society. With these considerations in mind, the International Conference on Water, Environment, Energy and Society (WEES-2016) was organized at AISECT University in Bhopal, MP, India, during March 15–18, 2016. The conference was fifth in the series and had several objectives.

The first objective was to provide a forum to not only engineers, scientists, and researchers, but also practitioners, planners, managers, administrators, and policy-makers from around the world for discussion of problems pertaining to water, environment, and energy that are vital for the sustenance and development of society.

Second, the Government of India has embarked upon two large projects: one on cleaning of River Ganga and the other on cleaning River Yamuna. Further, it is allocating large funds for irrigation projects with the aim to bring sufficient good-quality water to all farmers. These are huge ambitious projects and require consideration of all aspects of water, environment, and energy as well as society, including economics, culture, religion, politics, administration, law, and so on.

Third, when water resources projects are developed, it is important to ensure that these projects achieve their intended objectives without causing deleterious environmental consequences, such as water logging, salinization, loss of wetlands, sedimentation of reservoirs, loss of biodiversity, etc.

Fourth, the combination of rising demand for water and increasing concern for environmental quality compels that water resources projects are planned, designed, executed and managed, keeping changing conditions in mind, especially climate change and social and economic changes.

Fifth, water resources projects are investment intensive and it is therefore important to take a stock of how the built projects have fared and the lessons that can be learnt so that future projects are even better. This requires an open and frank discussion among all sectors and stakeholders.

Sixth, we wanted to reinforce that water, environment, energy, and society constitute a continuum and water is central to this continuum. Water resources projects are therefore inherently interdisciplinary and must be so dealt with.

Seventh, a conference like this offers an opportunity to renew old friendships and make new ones, exchange ideas and experiences, develop collaborations, and enrich ourselves both socially and intellectually. We have much to learn from each other.

Now the question may be: Why India and why Bhopal? India has had a long tradition of excellence spanning several millennia in the construction of water resources projects. Because of her vast size, high climatic variability encompassing six seasons, extreme landscape variability from flat plains to the highest mountains in the world, and large river systems, India offers a rich natural laboratory for water resources investigations.

India is a vast country, full of contrasts. She is diverse yet harmonious, mysterious yet charming, old yet beautiful, ancient yet modern. Nowhere we can find mountains as high as the snow-capped Himalayas in the north, the confluence of three seas and large temples in the south, long and fine sand beaches in the east as well as architectural gems in the west. The entire country is dotted with unsurpassable monuments, temples, mosques, palaces, and forts and fortresses that offer a glimpse of India's past and present.

Bhopal is located in almost the center of India and is situated between Narmada River and Betwa River. It is a capital of Madhya Pradesh and has a rich, several century-long history. It is a fascinating amalgam of scenic beauty, old historic city, and modern urban planning. All things considered, the venue of the conference could not have been better.

We received an overwhelming response to our call for papers. The number of abstracts received exceeded 450. Each abstract was reviewed and about two thirds of them, deemed appropriate to the theme of the conference, were selected. This led to the submission of about 300 full-length papers. The subject matter of the papers was divided into more than 40 topics, encompassing virtually all major aspects of water and environment as well energy. Each topic comprised a number of contributed papers and in some cases state-of-the-art papers. These papers provided a natural blend to reflect a coherent body of knowledge on that topic.

The papers contained in this volume, "Energy and Environment," represent one part of the conference proceedings. The other parts are embodied in six companion volumes entitled, "Hydrologic Modelling," "Groundwater," "Environmental Pollution," "Water Quality Management," "Climate Change Impacts," and "Water Resources Management." Arrangement of contributions in these seven books was a natural consequence of the diversity of papers presented at the conference and the topics covered. These books can be treated almost independently, although significant interconnectedness exists among them.

This volume contains seven parts organized under two sections. The first section deals with environment containing four parts, whereas the second section, containing three parts, is on energy. Part I deals with some aspects of hydrologic impacts of global warming and anthropogenic changes. Part II is on bio-environment and discusses plants, biomass, and bacterial species. Part III focuses on chemical environment. Section one is concluded with Part IV on social environment. Section two starts out with Part V on solar energy. Hydropower is discussed in Part VI. The concluding Part VII deals with biogas.

The book will be of interest to researchers and practitioners in the field of water resources, hydrology, environmental resources, agricultural engineering, watershed management, earth sciences, as well as those engaged in natural resources planning and management. Graduate students and those wishing to conduct further research in water and environment and their development and management may find the book to be of value.

WEES-16 attracted a large number of nationally and internationally well-known people who have long been at the forefront of environmental and water resources education, research, teaching, planning, development, management, and practice. It is hoped that long and productive personal associations and friendships will be developed as a result of this conference.

College Station, USA — Vijay P. Singh, Conference Chair
Bhopal, India — Shalini Yadav, Conference Organizing Secretary
Hazaribagh, India — Ram Narayan Yadava, Conference Co-Chair

Acknowledgements

We express our sincere gratitude to Shri Santosh Choubey, Chancellor, and Dr. V.K. Verma, Vice Chancellor, Board of Governing Body, and Board of Management of the AISECT University, Bhopal, India, for providing their continuous guidance and full organizational support in successfully organizing this international conference on Water, Environment, Energy and Society on the AISECT University campus in Bhopal, India.

We are also grateful to the Department of Biological and Agricultural Engineering, and Zachry Department of Civil Engineering, Texas A&M University, College Station, Texas, USA, and International Centre of Excellence in Water Management (ICE WaRM), Australia, for their institutional cooperation and support in organizing the ICWEES-2016.

We wish to take this opportunity to express our sincere appreciation to all the members of the Local Organization Committee for helping with transportation, lodging, food, and a whole host of other logistics. We must express our appreciation to the Members of Advisory Committee, Members of the National and International Technical Committees for sharing their pearls of wisdom with us during the course of the Conference.

Numerous other people contributed to the conference in one way or another, and lack of space does not allow us to list all of them here. We are also immensely grateful to all the invited Keynote Speakers, and Directors/Heads of Institutions for supporting and permitting research scholars, scientists and faculty members from their organizations for delivering keynote lectures and participating in the conference, submitting and presenting technical papers. The success of the conference is the direct result of their collective efforts. The session chairmen and co-chairmen administered the sessions in a positive, constructive, and professional manner. We owe our deep gratitude to all of these individuals and their organizations.

We are thankful to Shri Amitabh Saxena, Pro-Vice Chancellor, Dr. Vijay Singh, Registrar, and Dr. Basant Singh, School of Engineering and Technology, AISECT University, who provided expertise that greatly helped with the conference organization. We are also thankful to all the Heads of other Schools, Faculty Members and Staff of the AISECT University for the highly appreciable assistance in different organizing committees of the conference. We also express our sincere thanks to all the reviewers at national and international levels who reviewed and moderated the papers submitted to the conference. Their constructive evaluation and suggestions improved the manuscripts significantly.

Sponsors and Co-sponsors

The International Conference on Water, Environment, Energy and Society was Jointly organized by the AISECT University, Bhopal (M.P.), India and Texas A&M University, Texas, USA in association with ICE WaRM, Adelaide, Australia. It was partially supported by the International Atomic Energy Agency (IAEA), Vienna, Austria; AISECT University, Bhopal; M.P. Council of Science and Technology (MPCOST); Environmental Planning and Coordination Organization (EPCO), Government of Madhya Pradesh; National Bank for Agriculture and Rural Development (NABARD), Mumbai; Maulana Azad National Institute of Technology (MANIT), Bhopal; and National Thermal Power Corporation (NTPC), Noida, India. We are grateful to all these sponsors for their cooperation and providing partial financial support that led to the grand success to the ICWEES-2016.

Contents

Part VII Energy: Biogas

About the Editors

Prof. Vijay P. Singh is University Distinguished Professor, Regents Professor, and the inaugural holder of the Caroline and William N. Lehrer Distinguished Chair in Water Engineering in the Department of Biological and Agricultural Engineering and Zachry Department of Civil Engineering at Texas A&M University. He received his B.S., M.S., Ph.D., and D.Sc. degrees in engineering. He is a registered professional engineer, a registered professional hydrologist, and an Honorary Diplomate of American Academy of Water Resources Engineers.

Professor Singh has extensively published the results of an extraordinary range of his scientific pursuits. He has published more than 900 journal articles; 25 textbooks; 60 edited reference books, including the massive Encyclopedia of Snow, Ice and Glaciers and Handbook of Applied Hydrology; 104 book chapters; 314 conference papers; and 72 technical reports in the areas of hydrology, ground water, hydraulics, irrigation engineering, environmental engineering, and water resources.

For his scientific contributions to the development and management of water resources and promoting the cause of their conservation and sustainable use, he has received more than 90 national and international awards and numerous honors, including the Arid Lands Hydraulic Engineering Award, Ven Te Chow Award, Richard R. Torrens Award, Norman Medal, and EWRI Lifetime Achievement Award, all given by American Society of Civil Engineers; Ray K. Linsley Award and Founder's Award, given by American Institute of Hydrology; Crystal Drop Award, given by International Water Resources Association; and Outstanding Distinguished Scientist Award given by Sigma Xi, among others. He has received three honorary doctorates. He is a Distinguished Member of ASCE, and a fellow of EWRI, AWRA, IWRS, ISAE, IASWC, and IE and holds membership in 16 additional professional associations. He is a fellow/member of 10 international science/engineering academies. He has served as President and Senior Vice President of the American Institute of Hydrology (AIH). Currently he is editor-in-chief of two book series and three journals and serves on editorial boards of 20 other journals.

Professor Singh has visited and delivered invited lectures in all most all parts of the world but just a sample: Switzerland, the Czech Republic, Hungary, Austria, India, Italy, France, England, China, Singapore, Brazil, and Australia.

Prof. Shalini Yadav is Professor and Head of the Department of Civil Engineering, AISECT University, Bhopal, India. Her research interests include solid and hazardous waste management, construction management, environmental quality and water resources. She has executed a variety of research projects/consultancy in Environmental and Water Science and Technology and has got rich experience in Planning, formulating, organizing, executing and management of R&D programs, seminars, and conferences at national and international levels. She has got to her credit guiding an appreciable number of M.Tech. and Ph.D. students. She has published more than 10 journal articles and 30 technical reports. Dr. Shalini has also visited and delivered invited lectures at different institutes/universities in India and abroad, such as Australia, South Korea, and Kenya.

Professor Shalini Yadav graduated with a B.Sc. in Science from the Bhopal University. She earned her M.Sc. in Applied Chemistry with a specialization in Environmental Science from Bhopal University and M.Tech. in Civil Engineering with a specialization in Environmental Engineering from Malaviya National Institute of Technology, Jaipur, India in 2000. Then she pursued the degree of Ph.D. in Civil Engineering from Rajiv Gandhi Technical University, Bhopal, India in 2011. Also, she is a recipient of national fellowships and awards. She is a reviewer for many international journals. She has been recognized for one and half decades of leadership in research, teaching, and service to the Environmental Engineering Profession.

Dr. Ram Narayan Yadava holds the position of Vice Chancellor of the AISECT University, Hazaribagh, Jharkhand. His research interests include solid mechanics, environmental quality and water resources, hydrologic modeling, environmental sciences and R&D planning and management. Yadava has executed a variety of research/consultancy projects in the area of water resources planning and management, environment, remote sensing, mathematical modeling, technology forecasting, etc.

He has got adequate experience in establishing institutes/organizations, planning, formulating, organizing, executing and management of R&D programs, seminars, symposia, conferences at national and international level. He has got to his credit guiding a number of M.Tech. and Ph.D. students in the area of mathematical sciences and Earth sciences. Dr. Yadava has visited and delivered invited lectures at different institutes/universities in India and abroad, such as USA, Canada, United Kingdom, Thailand, Germany, South Korea, Malaysia, Singapore, South Africa, Costa Rica, and Australia.

He earned an M.Sc. in Mathematics with a specialization in Special Functions and Relativity from Banaras Hindu University, India in 1970 and a Ph.D. in Mathematics with specialization in Fracture Mechanics from Indian Institute of Technology, Bombay, India, in 1975. Also, he is a recipient of Raman Research

Fellowship and other awards. Dr. Yadava has been recognized for three and half decades of leadership in research and service to the hydrologic and water resources profession. Dr. Yadava's contribution to the state of the art has been significant in many different specialty areas, including water resources management, environmental sciences, irrigation science, soil and water conservation engineering, and mathematical modeling. He has published more than 90 journal articles; 4 textbooks; 7 edited reference books.

Part I
Environment: Hydrologic Impacts

Structural Evaluation of Cell-Filled Pavement

Subrat Roy and K.K. Pathak

Abstract This paper describes the findings of a study carried out for evaluating the performance of cell-filled pavement for low-volume roads. Details of laboratory investigations and the methodology adopted for construction of cell-filled pavement are presented. The aim of this study is to evaluate the structural behavior of cement concrete filled cell pavement laid over three different types of subbases (Water-bound macadam, soil–cement and moorum). A formwork of cells of thin plastic sheet was used to construct the cell-filled pavements to form flexible, interlocked block pavements. Surface deflections were measured using falling weight deflectometer and benkelman beam methods. Resilient moduli of pavement layers were estimated from the measured deflections. A comparison of deflections obtained from both the methodologies is also presented.

Keywords Cell-filled pavement · Low-volume roads · Falling weight deflectometer · Moorum

Introduction

Many states in India face the problem of scarcity of road quality aggregates and sand. Cost of aggregates escalated due to different restrictions enforced on the quarrying industry. In many places, carriage cost of aggregates and sand is very high resulting in high cost of construction of roads. Conventional flexible pavements for village roads consist of granular subbase and base with a wearing course of thin bituminous layer. The thick layers of granular base and subbase use significant quantities of aggregates. Traditional flexible pavements require regular maintenance which has been found to be difficult in the absence of adequate funds

S. Roy (✉) · K.K. Pathak
Civil and Environmental Engineering Department, NITTTR, Bhopal, India
e-mail: sroy.civil@gmail.com

K.K. Pathak
e-mail: kkpathak@nitttrbpl.ac.in

V.P. Singh et al. (eds.), *Energy and Environment*, Water Science and Technology Library 80, https://doi.org/10.1007/978-981-10-5798-4_1

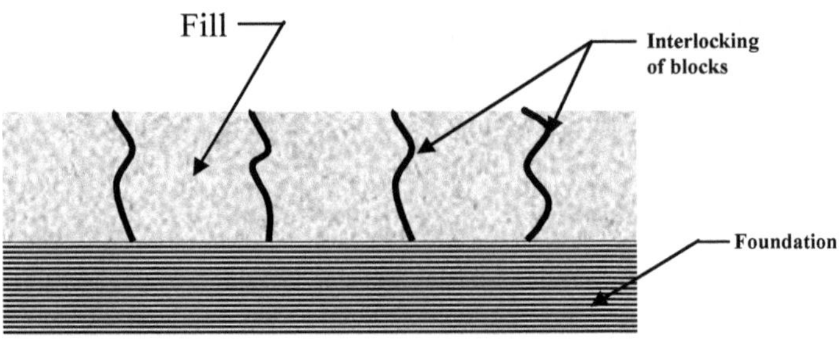

Fig. 1 Schematic representation of deformed plastic cells

and proper guidelines in the case of village roads. Therefore, it is necessary to explore alternative types of pavements that can be constructed using locally available materials and which require less frequent maintenance.

The concept of cell-filled pavement was developed in South Africa (Visser 1994; Visser and Hall 1999, 2003). Cell-filled pavement consists of a formwork of cells of plastic sheet laid over a compacted subbase. The form work of cells is stretched on the carriageway and held in position with the help of steel pegs driven into the base at suitable locations. Figure 1 shows the schematic arrangement of a formwork of cells stretched over a base. Cement-bound materials are filled into the cells and compacted. During compaction, cell walls get deformed and provide interlocking vertical joints between the blocks as illustrated in Fig. 1.

Due to the relatively larger flexibility of the cast in situ block pavements in comparison to rigid (concrete) pavements, these are termed as flexible-concrete pavements (Visser and Hall 1999). The present study aims at investigating the cell-filled pavement layers laid over different types of subbases.

Three test sections, each of 2.5 m × 2.5 m size, were constructed inside the IIT Kharagpur campus as per the following details.

- 100 mm thick roller compacted concrete filled cell pavement laid over 150 mm thick moorum subbase.
- 100 mm thick roller compacted concrete filled cell pavement laid over 150 mm thick water-bound macadam (WBM) subbase.
- 100 mm thick roller compacted concrete filled cell pavement laid over 150 mm thick soil–cement subbase.

Roller compacted concrete (RCC) is a stiff low water mix concrete compacted using a roller. Because of its low water cement ratio, RCC typically has high strength similar to or even greater than conventional concrete. Surface deflections of the test sections were measured using Falling Weight Deflectometer and Benkelman beam methods. Effective moduli of cell-filled pavement layers were back calculated. A comparison of deflections obtained from both the methodology is also presented.

Material Used for Construction of Pavements

Plastic Cell

Polyethylene sheet of 200 μm thickness, readily available in the market, was used for preparing the formwork of cells of 150 mm × 150 mm size. Depth of formwork was maintained as 100 mm in accordance with the proposed thickness of cell-filled layer. Strips of polyethylene sheets were heat-welded at appropriate locations to fabricate the formwork of cells.

Cement

Portland slag cement meeting the requirement of IS 455 (1989) was used for preparing the cement concrete and for stabilizing soil.

Soil and Moorum

Properties of the soil and moorum used are presented in Table 1.

Crushed Aggregate for Surface Layer

Crushed aggregates collected from Rampurhat in the state of West Bengal were used for preparing cement concrete. The coarse and fine aggregates were blended and the combined gradation adopted for roller compacted concrete (RCC) is shown in Table 2 (MORD 2004).

Table 1 Properties of soil and moorum

S. No.	Property	Value		Procedure followed
		Soil	Moorum	
1	Optimum moisture content	13.8%	10.3%	IS 2720—Part 7 (1980)
2	Maximum dry density	1940 kg/m^3	2136 kg/m^3	IS 2720—Part 7 (1980)
3	CBR			IS 2720—Part 16 (1987)
	Soaked (4 day)	7.8%	21.87%	
	Unsoaked	10.5%	28.6%	

Table 2 Aggregate gradation adopted for RCC

Sieve size (mm)	Percent passing by weight (%)
26.5	100
19.0	80–100
9.5	55–75
4.75	35–60
0.600	10–35
0.075	0–8

Crushed Aggregate for WBM Subbase

Crushed aggregates collected from Rampurhat in the state of West Bengal were used for water-bound macadam (WBM) subbase course. Gradations of coarse aggregates and screenings are given in Tables 3 and 4 respectively.

Soil–Cement Mix for Subbase Layer

Unconfined compressive strength and durability were the two criteria considered for the design of soil–cement mix. Unconfined compressive strength test was conducted under static axial loading condition on cylindrical specimens at a constant strain rate of 1.25 mm/min as per IS 4332-Part 5 (1970). Results of unconfined compression strength tests are given in Table 5.

To evaluate the durability of soil–cement mix, wetting and drying test was conducted on cylindrical specimens of 50 mm diameter and 100 mm height prepared using 5, 10, 15% cement contents and cured for 28 days. The test was performed according to ASTM D599 (2003). The specimens with 5% cement content crumbled in the first cycle. However, the specimens with 10 and 15%

Table 3 Aggregate gradation adopted for WBM base

Sieve size (mm)	Percent passing by weight (%)
63	100
53	95–100
45	85–90
22.4	0–10
11.2	0–5

Table 4 Gradation for screenings

Sieve size (mm)	Percent passing by weight (%)
11.2	100
5.6	90−100
0.180	15–35

Table 5 Unconfined compressive strength test results

Cement content (%)	Curing period (days)	Average unconfined compressive strength (mPa)
5	7	2.1
5	28	3.0
10	7	3.8
10	28	6.0
15	7	4.5
15	28	8.1

cement contents showed good results. The average material loss of the specimens with 10 and 15% cement content was 0.8 and 0.4% respectively after 12 cycles.

The British requirement for soil–cement stipulates minimum unconfined compression strength of 2.8 mPa to cater to heavy traffic requirement (Noor Megat 1994). Canadian Portland Cement Association recommended a maximum allowable loss of 7% for clay soil and 10% for silt soil (Noor Megat 1994). For the present investigation, soil–cement mix with 10% cement content was considered as it satisfied both the strength and durability criteria.

Mix Composition and Strength of Cement Concrete

Roller Compacted Concrete

A minimum compressive strength of 30 mPa is desirable for concrete block for village roads (MORD 2004). Cement, sand and coarse aggregates were taken in the proportion of 1:1.25:2.5 by volume since volume batching is convenient in remote villages. The optimum moisture content for concrete mix was obtained as five percent by weight of total weight of coarse aggregates, fine aggregates and cement (MORD 2004). Concrete cubes of dimension 150 mm × 150 mm × 150 mm were cast and their 28-day compressive strength was determined in the laboratory. Average 28 day compressive strength was obtained as 38.6 mPa.

Construction of Test Sections

Three test sections with different base materials were constructed in front of Vikram Sarabhai Complex inside Indian Institute of Technology, Kharagpur campus. Length and width of each section are 2.5 and 2.5 m, respectively.

Preparation of Subgrade

One of the earthen shoulders of the service road in front of Vikram Sarabhai Complex was excavated to prepare the test sections. 7.5 m long, 2.5 m wide trench was excavated for this purpose. Average depth of excavation was 300 mm. The trench was backfilled with selected soil and was compacted at optimum moisture content by 80 kN static roller in two layers of 150 mm thickness each. The field density of subgrade soil was determined by core cutter method as per IS 2720—Part 24 (1975). Cores were taken from three locations. Average field dry density and moisture content of subgrade were found to be 1830 kg/m^3 and 11.4% respectively. The degree of compaction achieved was close to 95% of standard proctor compaction (1940 kg/m^3).

Construction of Subbase

Local soil was used for preparation of soil–cement subbase. 10% Cement was added to soil and mixed at optimum moisture content. After mixing, it was spread on the prepared subgrade to the required depth and rolled with 80 kN static roller. The compacted surface was kept under moist condition for 14 days using wet jute gunny mats. WBM subbase was laid on the prepared subgrade to required thickness and rolled with 80 kN static roller. MORTH procedure and specifications were adopted for the construction of WBM base. Moorum was laid on the subgrade to the required thickness and rolled with 80 kN static roller. Optimum moisture content was used for preparing moorum subbase.

Construction of Surface Layer

Formwork of plastic cells of size 2.5 m × 2.5 m was stretched over the carriage way. Steel pegs and clips were used to hold the cell walls vertical. Test sections were constructed using roller compacted concrete. Cement concrete, prepared in a concrete mixture was placed manually inside the plastic cells. Compaction of the layer was done using 80 kN static roller. Curing was done for 14 days using wet jute gunny mats.

Structural Evaluation of Cell-Filled Pavement

Using Falling Weight Deflectometer

The three pavement sections were structurally evaluated using a Falling Weight Deflectometer developed by IIT Kharagpur. FWD was positioned in such a way that the loading plate is in the interior of the 2.5 m × 2.5 m test section. Deflections were measured at five radial distances (0, 300, 600, 900 and 1200 mm) from the center of the pavement section. Deflections were normalized to a load of 41 kN. Deflection data obtained for the three pavement sections are given in Table 6.

Elastic moduli of pavement layers were estimated from the deflection data given in Table 6 using BACKGA, a genetic algorithm based back calculation software developed at IIT Kharagpur (Reddy et al. 2002). The program uses linear elastic layered theory for computation of surface deflections. Layer thicknesses, surface deflections measured at radial distances of 0, 300, 600, 900 and 1200 mm, Poisson ratio values of the pavement layers and loading details were the main inputs for BACKGA. The following GA parameters were used for the analysis.

Population size = 60, Maximum number of generations = 60, Probability of crossover = 0.74, Probability of mutation = 0.1.

The back calculated moduli are given in Table 7.

Using Benkelman Beam

The pavement sections were also evaluated using Benkelman beam method. This was done for the purpose of comparing the deflection data obtained from FWD with those obtained using Benkelman beam which is a relatively simple procedure commonly adopted in India for structural evaluation of pavements. The probe of the beam was positioned in the middle of the 2.5 m × 2.5 m pavement section. A truck

Table 6 Measured deflection data from FWD

S. No.	Layer combinations	Deflection measurement (mm) at a radial distance (mm) of				
		0	300	600	900	1200
1	100 mm RCC 150 mm M	0.5940	0.3009	0.1568	0.0721	0.0433
2	100 mm RCC 150 mm WBM	0.5636	0.3561	0.2063	0.1021	0.0665
3	100 mm RCC 150 mm SC	0.3944	0.2454	0.1920	0.1140	0.0685

RCC Roller compacted concrete, *SC* soil–cement, *M* moorums
WBM Water-bound macadam

Table 7 Backcalculated layer moduli

S. No.	Layer combinations	Elastic modulus (mPa)		
		Cell-filled layer	Base	Subgrade
1	100 mm RCC 150 mm M	3025	152	95
2	100 mm RCC 150 mm WBM	2501	159	101
3	100 mm RCC 150 mm SC	4077	660	91

loaded with bricks with a rear axle load of 82 kN was used for deflection survey. The truck was made to move away from its initial position at creep speed and the rebound deflection was measured at 300 mm intervals. Different load positions of dual wheel considered in the investigation are shown in Fig. 2. All the dial gauge readings were taken with the probe of the beam placed at point A. When the wheel was at position A, initial reading was taken. When the wheel moved to subsequent points, corresponding readings were noted. Near-complete rebound of deflection was obtained when the load moved to a distance of 1.2 m (position E). Shape of the deflection bowl was deduced from these readings. Deflections obtained for different test sections as per this procedure are given in Table 8.

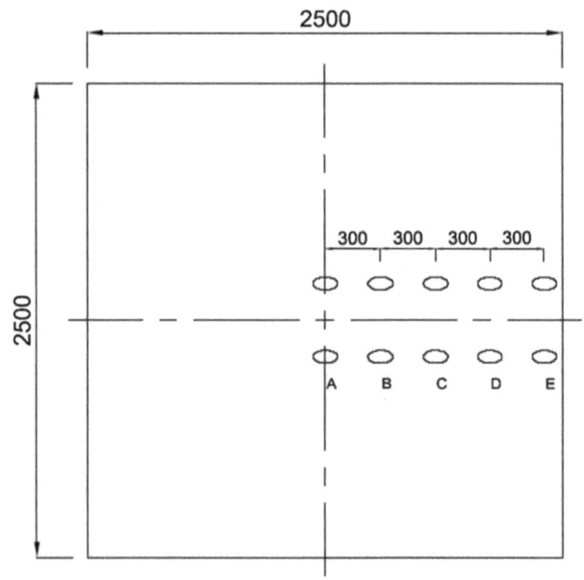

Fig. 2 Different load positions considered for deflection measurement

Table 8 Measured deflection data from Benkelman beam

S. No.	Layer combinations	Deflection measurement (mm) at a radial distance (mm) of				
		0	300	600	900	1200
1	100 mm RCC 150 mm M	0.8382	0.6604	0.3725	0.1863	0.0623
2	100 mm RCC 150 mm WBM	0.6953	0.3937	0.2286	0.1143	0.0781
3	100 mm RCC 150 mm SC	0.5956	0.2616	0.2008	0.1208	0.0772

Comparison of Deflections Measured Using Benkelman Beam Method and FWD

Comparison of deflections measured using Benkelman beam method and FWD on pavement sections is presented in Fig. 3. It can be observed from the figures that for most of the pavement sections, deflections obtained using Benkelman beam method is larger than those obtained using FWD. It is mainly due to the lower moduli of pavement layers under static/creep loading condition adopted in Benkelman beam method.

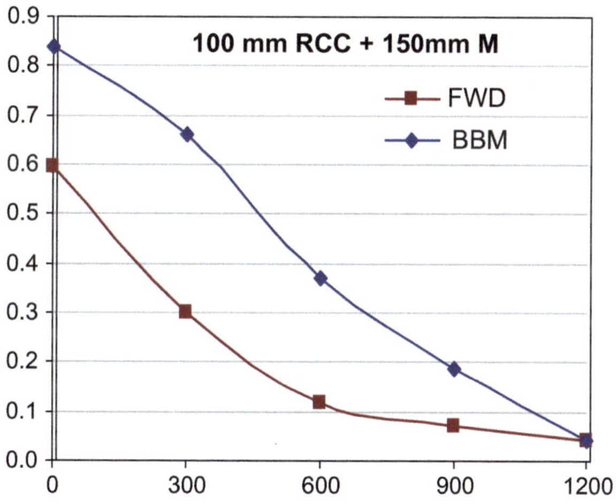

Fig. 3 Comparison of deflections radial distance (mm)

Conclusions

(a) Effective modulus of cement concrete (cell-filled) layer was about 4077 mPa when laid over soil–cement subbase. The corresponding values for moorum and WBM subbase were about 3025 and 2501 mPa respectively. Results indicate that the effective modulus of cell-filled layer depends on the strength of the subbase layer. Soil–cement subbase layer modulus was nearly 700 mPa whereas that of moorum and WBM subbase was 152 and 159 mPa respectively.
(b) Cell-filled pavements laid over different subbases have sufficiently high elastic modulus and they are expected to give strong pavement surfacing, free from reveling and pot holes.
(c) Because of its high modulus value, cell-filled pavement can be used as lower layer of high volume roads.
(d) In locations having scarcity of aggregates, cell-filled pavement laid over soil–cement and moorum subbase can be economically used.
(e) Recycled plastics can be used for fabricating the form work of cells. This will practically solve the disposal problem of high density polyethylene sheets.
(f) Making of form work of cells as well as construction of pavement are labor-intensive with minimal use of machinery. This technology will help employment generation in rural areas.
(g) Deflections obtained using Benkelman beam method are larger than those obtained using FWD.

References

ASTM D 599 (2003) Standard test methods for wetting and drying compacted soil-cement mixtures. American Society for Testing and Materials, United States

IS 4332—Part 5 (1970) Methods of test for soils—determination of unconfined compressive strength of stabilized soils. Bureau of Indian Standards, New Delhi, India

IS 2720—Part 24 (1975) Methods of test for soils—determination of dry density of soils in-place by core cutter method. Bureau of Indian Standards, New Delhi, India

IS 2720—Part 7 (1980) Determination of water content-dry unit weight relation using light compaction. Bureau of Indian Standards, New Delhi, India

IS 2720—Part 16 (1987) Methods of test for soils—laboratory determination of CBR. Bureau of Indian Standards, New Delhi, India

IS 455 (1989) Specification for Portland slag cement. Bureau of Indian Standards, New Delhi, India

MORD (2004) Specifications for rural roads. Ministry of Rural Development, Indian Roads Congress, New Delhi, India

Noor MJ (1994) Durability and strength characteristics of cement stabilized modified Melaka series. J Islamic Acad sci 7(2):137–141

Reddy MA, Reddy KS, Pandey BB (2002) Evaluation of effective pavement layer moduli using genetic algorithms. Int J Pavement Eng Asphalt Technol UK 3(1):6–19

Visser AT (1994) A cast in-situ block pavement for labour-enhanced construction. Concr Beton, Concr Soc S Afr 17:1–8
Visser TA, Hall S (1999) Flexible portland cement concrete pavement for village roads. Transport Res Record, Transport Res Board, Washington D.C. 1652:121–127
Visser TA, Hall S (2003) Innovative and cost effective solutions for roads in rural areas and difficult terrain. Transport Res Record: J Transp Res Board, Washington D.C. 1819A:169–173

Global Warming Issues—Need for Sustainable Drainage System in Urban Areas—Green Construction Technologies

A.K. Shrivastava, Shalini Yadav, L.S. Yadav, Shabana Khan, A.R. Khan and Sunil Sharma

Abstract The increasing population pressure on land in urban areas also adding the pressure on available fresh water sources, urban green spaces, biodiversity, and also on air and eco system. Another important issue of concern is the safe disposal of the ever-growing magnitude of urban waste. These key issues are important to be re-thinked with basic urban infrastructure planning, housing, transport, system, health infrastructure, and other allied civic amenities, required. All these issues necessitate a realistic approach to environmental policy with the basic urban planning of town and cities. The urban population of India as per 2011 Census was 377.10 million, with an average growth of 31% over the last decade, and accounting for 31.16% of the country's total population. This is projected to be reach to 600 million by the year 2030[(1)]. Thus in, the developmental policies decisions shall be multifunctional and participatory, taking into account the, environmental and other related issues and concerns of citizens. India is facing a typical problem of ground water table falling at a fast rate due to reduced recharge, and unplanned water withdrawal for agriculture and industry. Permeable pavements, if adopted for walkways, parking lots, can help in improving recharging of rainwater. This paper discusses technical information on the application, and necessity of pervious concrete pavements in urban areas, along with many of the environmental and economic benefits of using pervious concrete.

Keywords Environmental · Civil engineering · Material management
Risk · Sustainability · Eco system · Climate

A.K. Shrivastava (✉) · S. Yadav · L.S. Yadav · S. Khan · A.R. Khan · S. Sharma
Department of Civil Engineering, AISECT University, Bhopal, MP, India
e-mail: shrivastav1964@gmail.com

V.P. Singh et al. (eds.), *Energy and Environment*, Water Science and Technology Library 80, https://doi.org/10.1007/978-981-10-5798-4_2

Introduction

As the urbanization is increasing all over the world, due to various human activities, the cumulative effect of all the negative impacts of urbanization like increased pollution, production of heat from various activities—like fossil fuel burning, power production and consumption, use of air conditioners and fuel engines etc. The physical and chemical properties of the atmosphere are affected and the covering of the soil surface are becoming more impervious due to concrete and other coverings. This manmade, phenomena is causing rise in atmospheric temperature in cities/urban areas, as compared to rural and forest areas, and is named as UHI (Urban Heat Island) effect. The UHI adversely impacts not only residents of urban-related environments, but also other species and their associated ecosystem of the region.

In fact, UHI phenomena are indirectly related to climate change, in urban areas, due to its contribution to the greenhouse effect, and therefore, leads to the global warming phenomena. The UHI is also a 'negative type' modification of the climate, caused by changes to the form and composition of the land surface and gaseous composition in atmospheric surface. When a land cover of buildings and roads construction replaces the green spaces, the thermal, radioactive, moisture retention and aerodynamic properties of the land surface and the atmosphere are gets altered adversely. This causes rise in temperature nearby areas of buildings and other structures.

This growth of the importance of environmental sustainability in the world is underpinning effect of environmental, economic and societal matters of importance to engineers of the world over. Increasing urbanization and its changing structure has also resulted in drastic changes in our local and global environment. Estimates show that on an average towns/cities account for approx. 80% of energy consumption and approx. 75% of carbon emissions world over. Sustainable development is commonly defined as "meeting the needs of the present without compromising the ability of future generations to meet their own needs". The, green house gas (GHG) emissions, due to rapid industrialization is resulting global warming and other environmental disturbances all across the world. Hence the sustainable development concept has been gaining importance in construction industry also.

Considering the impact of environmental sustainability, issues on construction technologies, management, the solutions for the problem, related to increasing temperatures, due to rapid urbanization and industrialization is necessary. Our cities are being covered with buildings and air-proof concrete roads at a very rapid pace, thus making the environment uncomfortable and hot and also far away from natural climate, day by day. Because of the lack of water and air permeability due to common concrete and other pavements, the rainwater is not filtered to underground aquifers, which feed water to plants, throughout the year. Without constant supply of water to the soil, plants are difficult to grow normally. In addition, it is difficult for soil to exchange heat and moisture with air; therefore, the temperature and humidity of the Earth's surface in large cities cannot be controlled. This brings the phenomenon of hot islands in the city. At the same time, the splash on the road during rainy days reduces the safety of traffic of vehicles and footpath passengers.

The rapid urbanization has also resulted to weaken the linkages of people with nature and surroundings, and people are failing to recognize, the need to protect the environment. This has resulted in environmental degradation, particularly in cities, by increased air and water pollution, and problems of waste disposal and its management. However, restricting process of urbanization is not a solution to the problem; rather, it is important to ensure that it should proceeds in the right direction, with minimum or no negative impacts on the environment. Citizens of the cities are to be made responsible for minimizing the impact on the environment and to promote a sustainable pattern of living through their habits, behavior, and actions.

An urban area provides ample opportunities for better environmental management by the citizens through measures such as water conservation in homes and localities, recycling of waste, and expansion of public transport. This calls for a more holistic and innovative approach to environmental policy making and implementation. The multidimensional impact of climate change on the world is being studied in detail by the Intergovernmental Panel on Climate Change (IPCC), and is also being discussed at annual meetings of the Conference of Parties to the UN. Framework Convention on Climate Change (UNFCCC) adopted at Rio de Janeiro in 1992(2).

The year 2012 marked the 20th anniversary of the Rio Conference and extensive discussions were held at the Rio+20 Conference, on the progress made in the mitigation of climate change as well as adaptation to the consequences of climate change, such as increases in mean temperatures, adverse changes in precipitation, sea level rise, occurrence of extreme climatic events like drought, floods and coastal storms. The present trends in global warming can be controlled only by human action and intervention that can help to alleviate the adverse impacts of climate change. The Principle 10 of the Rio Declaration on Environment and Development states, "environmental issues are best handled with participation of all concerned citizens, at the relevant level". In India very little efforts have been made due to variety of reasons including poor strength of pervious concrete.

Still traditional pavements are being constructed along city roads, adding impervious surfaces in urban areas every day. The Pervious concrete methodology is now widely recognized as a sustainable construction material, as it helps to filter the runoff into ground, recharge groundwater aquifers, reduces storm water runoff quantity on roads and sewerage system, and can reduce the impact of the UHI effect, to certain extent. There is wide scope for research on use of pervious concrete, practical design for optimum strength, etc., to meet out requirement as environment friendly paving material for urban areas of the country, as well as all along highways.

Indian Climate and Monsoon

Monsoon rains are the backbone of Indian economy as most of our agricultural activities, are dependent on monsoon rains. The rivers flow and replenishment of ground water sources have a direct dependence on monsoon rains. Monsoon rains are a manifestation of the complex interactions between land, ocean and atmosphere. Urban areas also tend to be drier than their rural counterparts because of the lack of forests, trees, green spaces, and also predominance of impervious surfaces and urban drainage systems, which quickly remove water from the urban surface. This combination of these effects alters the energy balance of the earth, particularly the urban environment.

Consequently in urban area as, compared to rural areas, more of the sun's energy absorbed at the surface goes into heating the atmosphere and thus raising the air temperature area, resulting in a well-defined, distinct "warm island" among the "cool sea" represented by the lower temperature of the area's nearby natural landscape. Cities and urban areas are 3–8 °F (2–4 °C) warmer than surrounding areas due to the heat, island effect as shown in Fig. 1.

In India, an increase in the surface air temperature has been observed in the past century. A warming trend is visible along the west coast, central India, interior peninsula and the northeastern India, but some cooling trends are also visible in the northwest India and parts of south India (NAPCC-2008)[(3)].

It has been largely demonstrated that cities with variable landscapes and climates can exhibit temperatures several degrees higher than their rural surroundings areas (i.e., UHI effect). This phenomenon if increases in the future may result in a doubling of the urban to rural thermal ratio in the coming decades. Hence, proper assessment of the UHI—effect and strategies to control this effect, are becoming increasingly important for all government agencies and researchers of many

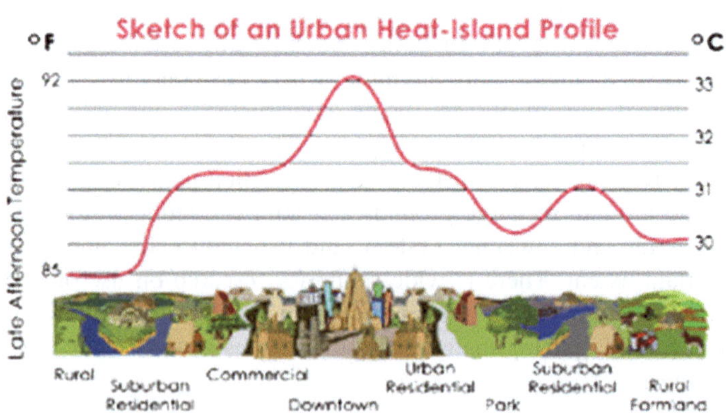

Fig. 1 Variation in temperature due to urban heat island effect. *Courtesy* Urban Heat Island Reduction, SRI, Portland Cement Association, www.ConcreteThinker.com 11/29/2015

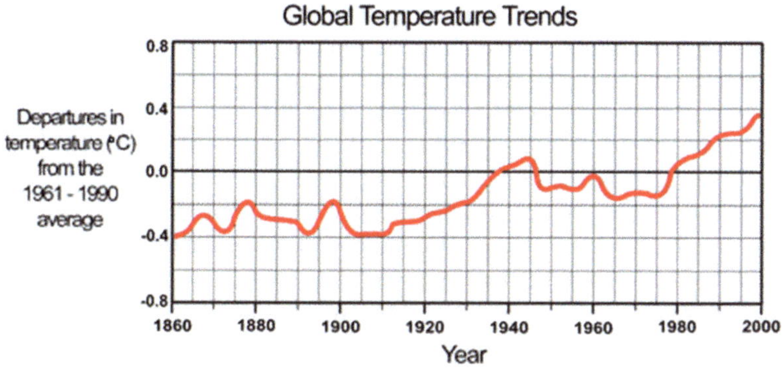

Fig. 2 *Image Courtesy* IPCC report on climate change 2001

affected countries. The climate change, by rise in general temperature, is recognized both as a threat and challenge to the environment of the world.

Climate has a significant role in the economic development of India. Since many sectors of the economy are climate sensitive. The tide gauge observations in the last four decades across the coast of India also indicate a rise in sea level at the rate of 1.06–1.25 mm/year. Further, some preliminary assessments point towards a warmer climate in the future over India, with temperatures, projected to rise by 2–4 °C by the year 2050s(3).

All the countries across the world are sincerely working out the impacts and associated vulnerabilities of their economies to impending projected climate change. However, there is no change in total quantity of rainfall is expected, however, spatial pattern of the rainfall are likely to change, with rise in number and intensity of extreme rainfall events. Figure 2 shows average rise of temperature during the last decade(4).

The continuous rise of general temperature and the changing rainfall pattern over the Indian region may jeopardize India's development by adversely impacting the natural resources such as water resources, forests, coastal zones, and mountains on which more than 70% of the rural population is dependent.

Reasons of Global Warming

Natural surfaces like grass and trees absorb sunlight and through photosynthesis they remove those global warming gasses, carbon dioxide, as well as other pollutants from the air. Also their roots and foliage help to trap storm water and move it to the underground aquifers. As asphalt pavements absorb more of the heat of the sun. This adds to the problem, day by day due to increase of road surfaces.

Because, asphalt is so widely used it creates heat islands as its black or gray surface absorbs sunlight. In summer its temperature may climb to 140 °F or more.

This heat requires more air conditioning thus requiring more electricity. Mostly electricity is generated using fossil fuels. So not only does this wastes money, but in making electricity, power generation plants create more greenhouse gasses. Even the impervious concrete absorbs much heat from the sun.

Of the above three, surfaces pervious concrete is the coolest one. When a rain falls on hot pavements, the rain evaporates more quickly, and less water reaches to the aquifer/other water bodies. It affects everything living in those waters bodies, changing the balance of nature. As more and more lands being covered by impervious surfaces, during rainy season, the storm water floods, the adjacent properties, roads, streams and destroying embankments, plant life and causing flooding in downstream areas, sometimes with resulting loss of lives also. Stream embankment gets eroded and has to be repaired regularly. On the other hand, pervious concrete diverts the rain water, to the safe zone, into the soil mass system and the aquifers.

Literature Review

The management of environmental sustainability is nowadays finds increasing importance to the construction industry worldwide. One of the key issues with respect to sustainability is the impact of urbanization, which mostly results in the conversion of pervious spaces to areas of impervious (paved) surfaces, leading to wide a range of ecology and environment related problems. As cities develop, countries grow in size and population, the urbanization with structures, roads, and industry grows as well. These significant changes create a series of problems related to storm water management; the most common issue is controlling the water volume. There are three main consequences of urbanization to storm water management:

(i) Alteration of hydrologic balance due to a lower water infiltration in ground.
(ii) An increased flow rate leading to flooding.
(iii) Storm water quality deterioration during runoff.

To find out the solutions of the problem, the research on pervious pavement materials has be started in the US and Japan in early 1980s. Pervious concrete is also widely used in Europe and Japan for roadway applications as a surface course to improve skid resistance and reduce traffic noise. However, the strength of the material is relatively low because of its porosity. The present construction practices adopted in many countries worldwide for sustainable roads parking pavements etc. are given in following sections.

Permeable Interlocking Concrete Pavement

Permeable interlocking concrete pavement (PICP) consists of manufactured concrete units that reduce storm water, runoff volume, and pollutants. The impervious units are designed in such a way to have, very small openings between the joints. The small openings typically comprise of approx., 5–15% of the paver surface area and are filled with, small-sized aggregates. The joints allow storm water to enter a crushed stone aggregate bedding layer and "open-graded" base i.e., crushed stone layers with no small or fine particles. The void spaces among the crushed stones stores water and infiltrate it, back into the soil mass system.

The stones in the joints provide 100% permeability and the base, filters storm water and reduces pollutants. The PICP openings provide high infiltration, having hydrological performance equal or better than other normal permeable surfaces. The small aggregate in the joints and bedding also facilitates interlock and load transfer to neighboring paver blocks. The pavers are available in many different shapes and sizes. Paving surface can be of different types depending upon the type of pavers. Figure 3 shows typical cross section of permeable pavement; Figs. 4 and 5, typical paving blocks and grid pavers, which can be utilized for pavements and parking areas.

Trees planted in parking lots and city sidewalks offer shade and produce a cooling effect in the area, further reducing heat island effects. Pervious concrete pavement is ideal for protecting trees in a paved environment (many plants have difficulty growing in areas covered by impervious pavements, sidewalks and landscaping, because air and water have difficulty getting to the roots). Pervious concrete pavements or sidewalks allow adjacent trees to receive more air and water

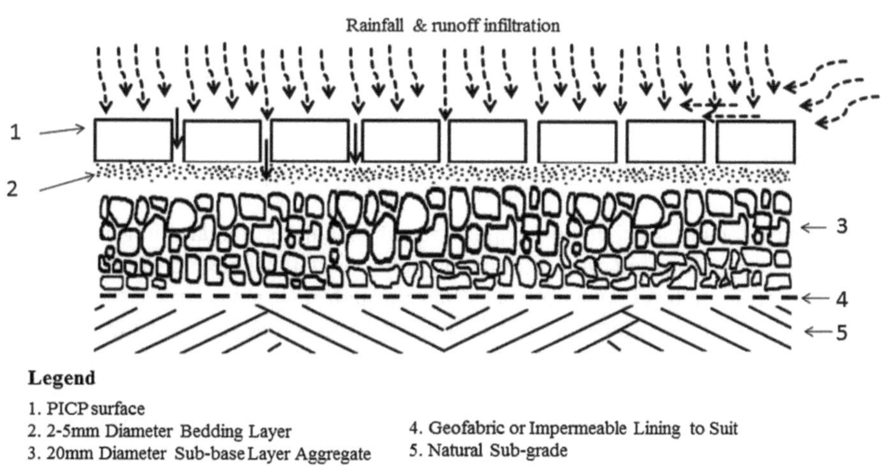

Fig. 3 Typical cross section of pervious surface

Fig. 4 Grid pavers for parking spaces. *Image Courtesy* google.co.in and Perviousconcrete.org

Fig. 5 Different arrangement of PICP pavements for parking. *Image Courtesy* google.co.in and perviousconcrete.org

and still permit, full use of the pavement. Pervious concrete provides a solution for landscapers and architects who wish to use greenery in parking lots and paved urban areas.

Permeable Asphalt

Permeable asphalt, also known or open-graded asphalt, is standard hot-mix asphalt with reduced proportion of sand/fine particles, and thus having slightly porous structure, which allows rain water to drain through it. It generally consists of fine and coarse aggregate stone particles bound by a bituminous-based binder coarse and laid in layers. Porous asphalt, laid over an aggregate storage bed reduces storm water runoff, rate, and pollutants, splashing over roads, reaching to sewerage system, rivers, and water bodies. The interconnected void space allows the rain water to flow through the asphalt and enter a bedding layer and supporting base, providing storage and runoff treatment. This water finally reaches to underground soil mass system, aquifers, and utilized by plants. Figure 6 shows permeable asphalts

Fig. 6 Permeable asphalts. *Image Courtesy* perviouspavements.org

having high drainage properties, through which rain water can percolate to aquifers below natural ground level.

The void space can be increased typically up to 15–20% by adjusting the amount of fine aggregates in the mix. Thickness of the asphalt generally depends on the traffic load, to be pass over the road, generally ranges from 75 to 150 mm. Porous asphalt can replace traditional impervious pavement for most pedestrian and vehicular applications. Porous asphalt performs well in pedestrian walkways, sidewalks, driveways, parking lots, and low-volume roadways.

Permeable Concrete

Permeable concrete, also known as porous concrete, or gap-graded concrete, is concrete with reduced proportion of sand or fines, in such a way to obtain porosity in concrete, and thus allows water to drain through it. Pervious concrete pavement can be a very effective mean to address important environmental issues in urban areas and can support sustainable infrastructural growth by arresting the rain water and allowing it to drain into the ground, thereby reducing flooding of roads, storm water, flooding problems in downstream areas and can be a good instrumental in recharging groundwater. Due to the high void content, pervious concrete is also lightweight concrete having density in range of, 1600–1900 kg/m^3. The compressive strength of pervious concrete is limited since the void content is so high. However, compressive strengths of 3.5–27.5 MPa are typical and sufficient for many applications such as Parking lots, Driveways, Sidewalks/walkways; Streets/road shoulders other light traffic areas.

Pervious concrete is a durable material, likewise the ordinary concrete; properly designed and constructed streets and parking lots, with pervious concrete method, will last many years with minimal maintenance. Thus concrete, conventional or pervious, is widely recognized as the lowest life cycle cost option available for pavement material selection worldwide. Pervious concrete also contributes to

Fig. 7 *Image Courtesy* google.com and pervious concrete.com

enhanced air quality by lowering atmospheric heating through light color and low density, decreasing the heat island effect, which is characterized by an up to 4 °C average temperature increase between an urban area. Figure 7 shows pervious pavement and standard concrete surface. Other picture shows interlocked paving blocks, with permeable joints in between. The pervious concrete has high drainage properties, through which rain water can percolate to ground.

Concrete for Sustainable Development

As Green Building Materials use of permeable concrete pavements can address the environmental issues and supporting sustainable development in urban areas. Permeable pavements can facilitate the following important issues of environment friendly material:

- Biodegradation of oils from cars and trucks,
- Rainwater infiltration into natural ground,
- Decrease urban heating,
- Replenish groundwater, allow tree roots to breathe,
- Reduce flash flooding.

However, the long term behavior of permeable pavement is still not well understood, and further research is needed to carry out.

Thus, permeable concrete have more voids in the structure leading to higher water infiltration into ground and air exchange rates compared with conventional concrete pavements. However, a current constraint to the development of permeable pavements is their perceived lack of structural strength. This is caused due to the need for greater porosity. The pervious concrete is not only an important contributor to sustainable practice, but that also it can, with proper mix design and targeted use of admixtures, achieve reasonable strength for use as a pavement construction material. As structures, buildings are constructed, the land properties are converted from "natural to developed land", thus impervious area is added and remaining pervious areas may gets compacted by construction equipment traffic. These changes reduce the potential for rainfall to infiltrate into the ground and increase the post development volume of storm water runoff discharged from the construction site. In addition to an increase in runoff volume, human activities associated with development routinely increase the concentration of nutrients (nitrogen and phosphorus), suspended solids, heavy metals, and many other pollutants in storm water runoff, which contribute to pollution of water bodies.

However, a current constraint to the development of permeable pavements is their perceived lack of structural strength. This is caused mainly by the need for greater porosity for treatment purposes. However the pervious, Concrete is can be utilized as effective mean in storm water management and sustainable development in urban areas.

Discussion

Permeable concrete pavements have a number of advantages with respect to sustainable construction practices and its management. In particular, permeable concrete pavements have considerable potential to manage runoff from urban landscapes, treat through natural biological processes runoff water, and indirectly manage heat/temperature of surroundings, and facilitate the growth of trees and manage pollution. They also have the potential to reduce noise pollution resulting from the impact of tires and pavement. These advantages mean that they are able to facilitate sustainable construction processes, and so assist the world construction industry for better respond to global sustainability challenges.

One important construction management issue is the need to ensure quality of the finished product through careful concrete mix design and the construction process. This paper has addressed one of the aspects of this quality management process—the strength of permeable concrete pavement material. From the construction management point of view, disadvantages include the need to design the pavement, mix and construction process well to ensure maximum strength and bonding; and, because of their lower strength compared to other aggregates, to limit recycled aggregates to at best a base course material.

Apart from the need to undertake tightly specified and well quality controlled mix design and placement, and close attention to placement of materials used in permeable concrete pavements, a potential disadvantage with permeable concrete pavements is their ability to being able to manage clogging issues like muddy water. This potential disadvantage requires further research. In addition, given the risks previously identified in this paper with potential leaching from binder material in recycled concrete aggregate with respect to normal road pavements, further research in minimizing this risk is required. A diagrammatic comparison between

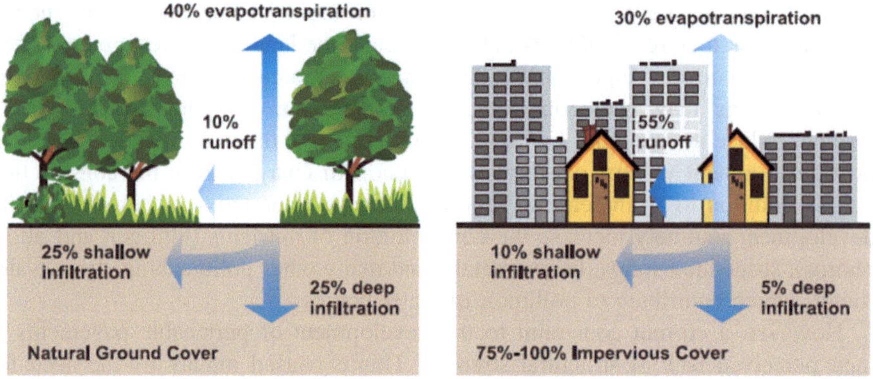

Fig. 8 *Image Courtsey* learnnc.org/lp/editions/mudcreek/6394

natural ground (pervious) and impervious after land covered by building and structures is shown in Fig. 8. In case of impervious layers, the runoff is around 55% as compared to 10% with natural ground cover.

Conclusion

The present paper has highlighted the importance of reducing environmental impact through use of pervious concrete pavements, for sustainability in the concrete industry for civil engineering students. It highlights the importance of reducing environmental, impact while maintaining low cost.

Through the urbanization process the natural environment get affected, since roads are constructed, particularly in urban areas, the UHI effect increases as more and more land is converted into developed one. The permeable concrete has good potential to make a positive contribution to sustainable road construction and life cycle management. It can meet stakeholder requirements through less impact on the environment on which roads are constructed, and therefore can assist the construction industry to move closer to sustainable construction management.

The major issue that needs attention is the need to closely apply quality management to pavement and mix design, and concrete placement. More research is required to better manage its disadvantages, such as the possible potential to clog under certain circumstances and to minimize any leaching effects into the environment from binder material.

Image Courtesy ;-MANDALA CONCRETE, LLC (504)430-1609 brentmagee6@gmail.com

Image courtesy:-Handbook for,Pervious Concrete Certification, in Greater Kansas City,www.concretepromotion.com

GO-GREEN FOR SUSTAINBLE DEVELOPMENT

References

Beanl EZ, Huntl WF, Bidelspach DA, Burak RJ (2004) Study on the surface infiltration rate of permeable pavements. In: 1st water and environment specialty conference of the canadian society for civil engineering, Saskatoon, Saskatchewan, Canada 2–5 June 2004

Brown HJ (2013) Pervious concrete research compilation: past, present and future updated, August 2013. Concrete Industry Management Program, Middle Tennessee State University

Environmental survey report-TERI-2014

Green highways: an initiative towards sustainable development press information Bureau Government of India, 28 Aug 2015

Gupta S (2010) Design of porus pavements; CE682 infrastructure and transport planning. Indian Institute of Technology, Kanpur, 6 Mar 2010

Harshavarthana Balaji M, Amarnaath MR, Kavin RA, Jaya Pradeep S (2015) Design of eco-friendly pervious concrete. Int J Civil Eng Technol (IJCIET) 6(2). ISSN 0976–6308 (Print), ISSN 0976–6316 (Online)

India: Greenhouse Gas Emissions 2007, Ministry of Environment and Forests Government of India, May 2010 (INCCA Indian Network for Climate Change Assessment). IPCC report-2001

Katarial BM, Pathak VB, Shah JA (2013) Sustainable development a rethinking of, construction industry in India. Int J Innovative Res Sci Eng Technol 2(10)

Khurd VG, Nitish M (2015) Experimental study of properties of pervious, concrete as a pavement material. Int J Adv Eng Nano Technol 2(9)

Nadgouda K, Ghadi P, Gharat M, Gurav R, Margaje T, Shirke E (2014) The use of pervious concrete in rainwater, management. Nirma Univ J Eng Technol 3(1)

Presentation on: Porous Pavement/Groundwater, Recharge 19382 Cape May, NJ 08204 Daniel Wible, P.E., L.E.E.D. A.P. Cahill Associates Environmental Consultants. www.thcahill.com

Ravindran KT. Chennai floods present a lesson in urban planning, Dec 25. Hindustan Times, Chennai

Shah DS, Pitroda J, Bhavsar JJ (2013) Pervious concrete: new era for rural road, pavement. Int J Eng Trends Technol (IJETT) 4(8) (Report of the RMC Research & Education Foundation)

Solis AV, Durham SA, Ramaswami A. Providing stormwater management solutions in Rajkot, India. A pervious concrete system demonstration. Int J Constr Environ 2(3):135–154

Sri Ravindrarajah SR (2010) Environmental friendly pervious concrete for sustainable construction. Nanyang Technological University, Singapore & University of Technology, Sydney,

Australia. In: Yukari A, 35th conference on our world in concrete and structures, University of Technology, Sydney, Australia, 25–27 Aug 2010, Singapore
Statistics Related to Climate Change in India, Ministry of Statistics and Programme Implementation Government of India, Central Statistical Office, Social Statistics Division, New Delhi, November, 2013. www.mospi.gov
Tayabji S (2015) New Delhi workshop—concrete pavements: current, developments. World Bank, Feb 2015.
Thorpe D, Zhuge Y (2010) Advantages and disadvantages in using permeable concrete pavement as a pavement construction material. In: Egbu C (ed) Proceedings of 26th annual ARCOM conference, 6–8 Sept 2010. Leeds, UK
Uma Maguesvaria M, Narasimhab VL (2013) Studies on characterization of pervious concrete for pavement, applications. In: 2nd conference of transportation research group of India (2nd CTRG) 1877-0428 © 2013. Elsevier

Part II
Environment: Bio-environment

Water Hyacinth: A Useful Plant to Improve Rural Economy

Priti Mathur and S.M. Mathur

Abstract Aquatic plants have created havoc all over the world. Natural aquatic plants population to some extent is healthy for any aquatic ecosystem as it serves as fish feed, plays an important role in nutrient cycling, purifies the water, controls unwanted algal growth and supports fauna including birds. Unwanted growth of aquatic plants especially some seriously invasive species like water hyacinth and *Salvinia* has caused a loss of billions of US$ all over the world. It affects the supply of drinking water and water hyacinth (*Eichhornia crassipes*) has proved to be a persistent and expensive aquatic weed problem costing millions of dollars to control and unaccounted millions of dollars more to the damage to the environment, irrigation systems and crops. It was first introduced as an ornamental plant in India in 1896 from Brazil. Water hyacinth has posed ecological and economical problems in several countries. This weed poses problem especially in tropical and subtropical countries where environmental conditions provide a year round growing period. The natural loss of water from the water surface by evaporation is thought to increase through transpiration from the leaves of water hyacinth by at least 40–50%. The dense growth of water hyacinth obstructs water flow in irrigation channels, interferes with navigation and hydroelectric power generation. The flow of water is reduced by 40–95% and roughness coefficient increases from 0.024 to 0.055 in irrigation channel. Water hyacinth interferes with the seed germination and seedling establishment in paddy, resulting in heavy economic losses to the tune of up to 24 million dollars. The oxygen-depleting load of one hectare of water hyacinth mat is equivalent to that of the sewage created by 80 persons. Over the years, various control methods have been studied and tried including chemical, biological and mechanical means but with no lasting success. The simplest way to control this

P. Mathur (✉)
College of Home Science, MPUAT, Udaipur 313001, Rajasthan, India
e-mail: pritimathur710@yahoo.co.uk

S.M. Mathur
Department of FM&PE, College of Technology and Engineering, MPUAT, Udaipur, India
e-mail: Shiloo2009@gmail.com

V.P. Singh et al. (eds.), *Energy and Environment*, Water Science and Technology Library 80, https://doi.org/10.1007/978-981-10-5798-4_3

weed is to harvest it and utilize it for useful products. Utilization of plant will improve rural economy and also solve the problem of management of this weed in India.

Keywords Aquatic weeds · Harvesting · Water · Fertilizer · Furniture Biogas

Introduction

Lakes are important natural resources. As source of recreation, they support fishing, boating, supply of drinking water, swimming, and aesthetic enjoyment. Excessive growth of water weeds namely water hyacinth, creates a large numbers of problems particularly related to the use and management of water resources. Water hyacinth (Fig. 1) has proved to be a persistent and expensive aquatic weed problem costing millions of dollars to control and unaccounted millions of dollars more, to the damages to the environment, irrigation systems and crops. It forms dense mats that interfere with navigation, recreation, irrigation, and power generation. These mats competitively exclude native submersed and floating-leaved plants. Low oxygen conditions develop beneath water hyacinth mats and the dense floating mats impede water flow and create good breeding conditions for mosquitoes. Water hyacinths are a severe environmental and economic problem in all of the gulf coast states and in many other areas of the world with a subtropical or tropical climate. This species has rapidly spread throughout inland and coastal freshwater bays, lakes, and marshes in the United States and in other countries. Water hyacinth reproduces sexually by seeds and vegetative by budding and stolen production. Daughter plants sprout from the stolons and doubling times have been reported of 6–18 days. The seeds can germinate in a few days or remain dormant for 15–20 years. Over the years various control methods have been studied and tried including chemical, biological and mechanical means but with no lasting success. Water hyacinth persists and continues to spread. The problem is often aggravated by the over increasing nutrient enrichment (eutrophication) of our water bodies brought about by excessive use of fertilizers, industrialisation and population growth simultaneous with the rising demand for clean water.

This weed is a problem especially in tropical and subtropical countries where environmental conditions provide a year round growing period. The natural loss of water from the water surface by evaporation is thought to increase through transpiration from the leaves of water hyacinth by at least 40–50% (Lindsey and Hirt 1999). Due to this sometimes it is called "shokh samunder" in India. The dense growth of water hyacinth obstructs water flow in irrigation channels, interferes with navigation and hydroelectric power generation. The flow of water is reduced by 40–95% and roughness coefficient increases from 0.024 to 0.055 in irrigation channel (Lindsey and Hirt 1999). Water hyacinth interferes with the seed germination and seedling establishment in paddy, resulting in heavy economic losses to the tune of

Fig. 1 Water hyacinth plants

up to 24 million dollars. The oxygen-depleting load of one hectare of water hyacinth mat is equivalent to that of the sewage created by 80 persons (Gopal and Sharma 1981). This action lowers the natural ability of water body to absorb organic pollution and create septic and odorous conditions. Besides, this increases carbon dioxide concentration, decreases dissolved oxygen content and impairs clarity of water.

Water Hyacinth Control and Its Applications

Chemical control using herbicides, the most popular of which 2,4-dichlorophenoxy acetic acid, has so far been the main weapon used in controlling water hyacinth. It is effective and relatively cheaper than the other mechanical and manual control methods. However, their long-term effects on the environment are unknown and the sprayed plants are left to rot in the water, causing pollution and aggravated eutrophication. Biological control, on the other hand, promises to be a cheaper control method in the long run because it is self-perpetuating and has a long-lasting effect (Julien et al. 2001).

Water hyacinth control by mechanical and manual harvesting has been widely practised in many countries (Fig. 2). The simplest mechanical method of control would be to chop the water hyacinth in situ and leave it to rot in the water. This, however, has two main drawbacks; the plant is wasted (not utilized) in the process of decay and the chopped plant extract oxygen from the water causing suffocation

Fig. 2 Manual removal of water hyacinth

to animal life in water body. Manual harvesting and use of simple tools are widely practised, particularly in developing countries but are generally inadequate to control hyacinth because they are very slow. Three hundred man-hours are required to remove 1 ha of water hyacinth (Lindsey and Hirt 1999). Use of power-operated machinery improves the performance over manual methods, but is still slow to cope with the explosive growth of hyacinth and is also expensive when compared with chemical control methods. Still, mechanical control of hyacinth is more popular and has important advantages over chemical and biological methods as it is environmentally safe and has the potential to control the currently serious and pervasive problem of eutrophication. The biggest problem with mechanical control of this weed is its transportation because of its bulk.

Mechanical removal of water hyacinth is seen as the best short-term solution to the proliferation of the plant. Mats of water hyacinth can be enormous and can have a density of up to 200 tonnes per acre. Manual removal of water hyacinth is suitable only for small areas. It is labour-intensive work and in some areas involves serious health risks associated with the work (crocodiles, hippopotamus and bilharzia in Lake Victoria for example). The cost of removal can be reduced by effective utilization of the plant with a variety of applications listed in following paragraphs.

Paper: The water hyacinth fibre alone does not make a particularly good paper but when the fibre is blended with waste paper or jute the result is good. The pulp is mixed with bleaching powder, calcium carbonate and sodium carbonate before being heated. Small-scale cottage industry papermaking projects have been successful in a number of countries, including the Philippines, Indonesia, and India.

Fibre board: Another application of water hyacinth is the production of fibre-boards for a variety of end users. The House and Building Research Institute in Dhaka has carried out experimental work on the production of fibre boards from water hyacinth fibre and other indigenous materials. The boards are floated in a vat on water and then finished in a hand press and hung to dry. The physical properties of the board are sufficiently good for use on indoor partition walls and ceilings.

Yarn and rope: The fibre from the stems of the water hyacinth plant can be used to make rope. The stalk from the plant is shredded lengthways to expose the fibres and then left to dry for several days. The rope making process is similar to that of jute rope. The finished rope is treated with sodium metabisulphite to prevent it from rotting. In Bangladesh, the rope is used by a local furniture manufacturer who winds the rope around a cane frame to produce an elegant finished product.

Basket work: In the Philippines water hyacinth is dried and used to make baskets and matting for domestic use. The key to a good product is to ensure that the stalks are properly dried before being used. If the stalks still contain moisture then this can cause the product to rot quite quickly. In India, water hyacinth is also used to produce similar goods for the tourist industry. Traditional basket making and weaving skills are used.

Charcoal briquetting: It is that a small-scale water hyacinth charcoal briquetting industry could have several beneficial aspects for the lakeside communities. The technical aspects are yet to be fully developed

Biogas production: Water hyacinth has to be pre-treated before entering a digester (macerated, chopped or beaten) to promote digestion and to remove air entrapped in the tissue of the plant which would cause it to float. To reduce the need for large volume digesters high rate digestion techniques have been employed. One such design has been tested in Bangladesh by a team from Warwick University, UK and the Housing and Building Research Institute, Dhaka, Bangladesh. The design was for a small (8.3 m^3) baffled reactor which was fed with juiced water hyacinth. The through flow of the reactor was 1.2 cubic metres per day. Other studies have been carried out, primarily in India, with quantities of up to 4000 L of gas per tonne of semi dried water hyacinth being produced with a methane content of up to 64% (Gopal 1987). Most of the experiments have used a mixture of animal waste and water hyacinth. There is still no firm consensus on the design of an appropriate water hyacinth biogas digester.

Water purification: Water hyacinth can be used to aid the process of water purification either for drinking water or for liquid effluent from sewage systems. In a drinking water treatment plant, the water hyacinth has been used as part of the pre-treatment purification step. Clean, healthy plants have been incorporated into water clarifiers and help with the removal of small flocs that remain after initial coagulation and floc removal or settling. The result is a significant decrease in turbidity due to the removal of flocs and also slight reduction in organic matter in the water. In sewage systems, the root structures of water hyacinth (and other aquatic plants) provide a suitable environment for aerobic bacteria to function.

Animal fodder: In China, pig farmers boil chopped water hyacinth with vegetable waste, rice bran, copra cake and salt to make a suitable feed. In Malaysia, fresh water hyacinth is cooked with rice bran and fishmeal and mixed with copra meal as feed for pigs, ducks and pond fish. Similar practices are commonly used in Indonesia, the Philippines and Thailand (Anon 1976). The high water and mineral content mean that it is not suited to all animals. The use of water hyacinth for animal feed in developing countries could help solve some of the nutritional problems that exist in these countries. Animal feed is often in short supply and although humans cannot eat water hyacinth directly, they can feed it to cattle and other animals which can convert the nutrient into useful food products for human consumption.

Fertilizers: Water hyacinth can be used on the land either as a green manure or as compost. As a green manure it can be ploughed into the ground or used as mulch. The plant is ideal for composting. Microbial decomposition breaks down the fats, lipids, proteins, sugars and starches. The mixture can be left in piles to compost; the warmer climate of tropical countries accelerates the process producing, rich pathogen free, compost which can be applied directly to the soil. In Sri Lanka water hyacinth is mixed with organic municipal waste, ash and soil, composted and sold to local farmers and market gardeners.

Fish feed: The Chinese grass carp is a fast growing fish which eats aquatic plants. It grows at a tremendous rate and reaches sizes of up to 32 kg (National Academy of Sciences 1979). It is an edible fish with a tasty white meat. It will eat submerged or floating plants and also bank grasses. The fish can be used for weed control and will eat up to 18–40% of its own body weight in a single day (Gopal 1987). Water hyacinth has also been used indirectly to feed fish. Dehydrated water hyacinth has been added to the diet of channel catfish fingerlings to increase their growth (Gopal 1987). It has also been noted that decay of water hyacinth after chemical control releases nutrients which promote the growth of phytoplankton with subsequent increases in fish yield (Gopal 1987).

Transportation of the harvested weed to the disposal site is costly, because it very voluminous and has high water content. Chopping and crushing can reduce the volume and surface water and utilization of it will reduce the cost of operation and improve the economy of the villagers. Therefore, a water hyacinth chopper cum crusher was developed at College of Technology and Engineering, Udaipur (India) and tested extensively(Mathur and pratap 2004; Mathur 2005a, b and 2006).

Practical Utilization of Water Hyacinth

Despite several efforts with different control measures and heavy expenditure, man has had hardly any success in controlling the weed. To attain a positive approach for the management and practical utilization of water hyacinth one must think positively about the plant. It is a resource which should be utilized. It has a beautiful flower, grows vigorously and abundantly to produce a large biomass, has leaves

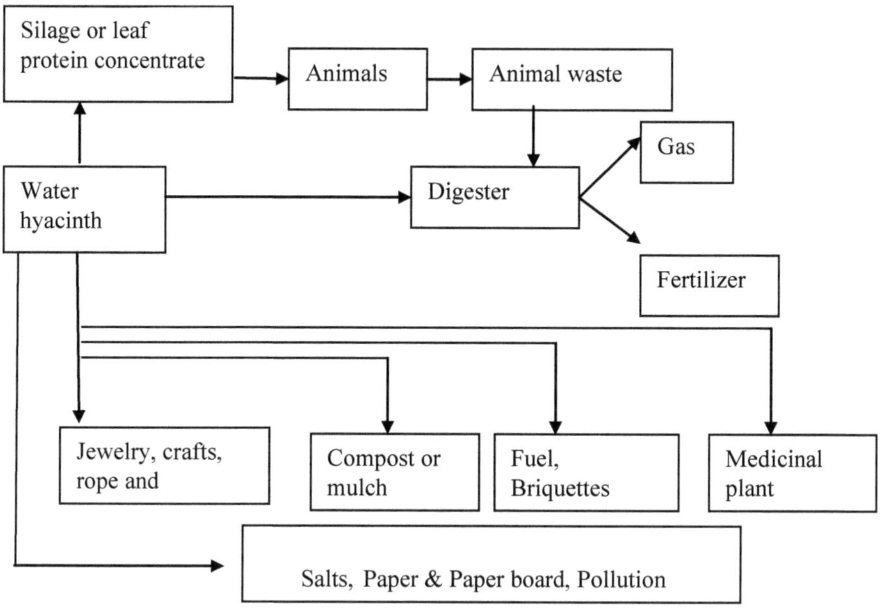

Fig. 3 Utilization of harvested chopped and crushed water hyacinth

rich in protein, has a fibrous stem, and has high potassium content and fibrous roots. The problem of management of water hyacinth is ideally solved by its utilization. Utilization has also been thought of as a means to reduce the cost of removal of the weed. Today, utilization is considered an important part of the weed management. In recent years hundreds of papers have appeared on the utilization of water hyacinth and the general trend of thinking leans towards management of water hyacinth by utilization. In particular, the following uses of water hyacinth have been demonstrated (Gopal and Sharma 1981; Aquaphyte 2001; Lindsey and Hirt 1999). Various applications of water hyacinth have been shown in Fig. 3.

References

Anonymous (1976) Making aquatic weeds useful: some perspectives for developing countries. National Acdemy of Sciences, Washington, DC

Aquaphyte (2001) Traditional medicinal knowledge about an obnoxious weed Jal Kumbhi *(Eichhornia crassipes)* in Chhattisgarh (India). A News Letter Published by Center for Aquatic and Invasive Plants, Gainesville, Florida, vol 21, no 2, p 18. ISSN-0893-7702

Gopal B (1987) Aqatic Plant Studies 1. Waterhyacinth. Elsevier Publishing, New York, USA

Gopal B, Sharma KP (1981) Water hyacinth. Hindustan Publishers, Delhi, India

Julien MH, Griffiths MW, Stanely JN (2001) Biological control of water hyacinth 2. Australian Centre for International Agricultural Research, Canberra, ACT 2601

Lindsey K, Hirt HM (1999) Use water hyacinth! A practical handbook of uses for the water hyacinth from across the world. Published by Drukerei Bauer, Winnenden, Germany
Mathur SM (2005a) Development and performance evaluation of water hyacinth chopper cum crusher. Report Published by the Department of Farm Machinery and Power Engineering, College of Technology and Engineering, Udaipur, India
Mathur SM (2005b) Key note paper on mechanical management of water hyacinth in fresh water lakes to boost lake environment. In: International proceedings on water and health, held at J S S Mahavidyapeetha, Mysore, 22–23 Jan, pp 26–30
Mathur SM (2006) Aquatic weeds and its impact on environment. Int J Ecol Environ Conser 3
Mathur SM, Pratap S (2004) Development and performance evaluation of water hyacinth chopper cum crusher. Int J Biosyst Eng UK 88(4):411–418
Yee AA and Associates, Inc., Suite 810, 1441 Kapiolani Boulevard, Honolulu, Hawaii

Growth, Biomass and Carbon Sequestration by Trees in Nutrient-Deficient Bhata Land Soil of Bilaspur, Chhattisgarh, India

K.K. Chandra and Atul Kumar Bhardwaj

Abstract Entisol soil is commonly known as bhata soil which covers 18.76% in Chhattisgarh state and 5.15% in Bilaspur district. The soil is red in colour, sandy texture with preponderances of hard spherical mass of iron oxides. This soil is often barren and is categorized as waste land due to its poor productivity, organic matter, water holding capacity, nutrient, microbial activities, and heavy biotic pressure. This wasteland needs to rejuvenate on sustainable management basis owing to mitigate the impact of climate change. The Forest Development Corporation has conducted the massive field plantations during 1990–91, using some forest tree species in this soil. One such plantation was established in 250 acre area at Chakarbhata, Bilaspur Chhattisgarh (India), where a pure block of 156 trees per species was accommodated in 0.25 ha at 4 × 4 m distance. The present investigation was conducted in 2014, and growth, biomass, and carbon sequestration by trees planted in bhata land were quantified to assess the magnitude and role of forests in relation to climate change. The tree height was recorded the highest in *Eucalyptus globulous* followed by *Albizia lebbeck* and lowest in *Pongamia pinnata*. Diameter was observed maximum of 43.12 cm in *A. lebbeck* and lowest 13.40 cm in *P. pinnata*. The total biomass was recorded maximum of 26.25 q $tree^{-1}$ in *A. lebbeck* and minimum 3.01 q $tree^{-1}$ in *P. pinnata*. The biomass accumulation in all tree species under study were *A. lebbeck* > *E. globulous* > *Azadirachta indica* > *Emblica officinalis* > *Terminalia arjuna* > *Peltophorum ferruginium* > *Dalbergia sissoo* > *Pongamia pinnata*. Similarly, the carbon sequestration potential was highest in *A. lebbeck* and lowest in *P. pinnata* in bhata land soil.

Keywords Entisol · Biomass · Carbon sequestration

K.K. Chandra (✉) · A.K. Bhardwaj
Department of Forestry, Guru Ghasidas Vishwavidyalaya, Bilaspur, Chhattisgarh, India
e-mail: kkckvk@gmail.com

V.P. Singh et al. (eds.), *Energy and Environment*, Water Science and Technology Library 80, https://doi.org/10.1007/978-981-10-5798-4_4

Introduction

Chhattisgarh is a power hub of India as more than 80 power plants are going to commission in the next few years. These units are utilizing coal for power generation and dumping CO_2 in the area. The situation will be more alarming because more and more industries are coming to the state due to surplus electricity and easily availability of manpower. To mitigate the problem of global warming, more and more area is to be afforested using highly potential tree species. Trees play a vital role in mitigating the diverse effects of environmental carbon degradation and increasing concentration of carbon dioxide in the atmosphere. Trees promote sequestration of carbon into soil and plant biomass. Therefore, tree-based land use practices could be viable alternatives to store atmospheric carbon dioxide due to their cost-effectiveness, high potential of carbon uptake, and associated environmental as well as social benefits (Dhruw et al. 2009). Increasing levels of carbon dioxide in the atmosphere during the past few decades have drawn the attention of the scientific community towards the process of carbon storage and soil organic carbon store. Carbon dioxide (CO_2) is a natural greenhouse gas in the atmosphere and is in part responsible for the earth's relatively stable climate. It is a "greenhouse" gas because it traps heat near the earth's surface, contributing to observed and predicted global warming. Human activities, especially the burning of fossil fuels such as coal and oil and destruction of natural forests, are greatly increasing the level of CO_2 in the atmosphere. Concentrations have risen from about 284 parts per million in 1832 to about 387 ppm in March 2009. Mean temperature increases between 1 and 6 °C have been projected over the next 70 years. The concentration of atmospheric carbon dioxide can be lowered either by reducing emissions or by enabling the storage of carbon dioxide in the terrestrial ecosystem (IPCC 2003). Trees remove carbon dioxide from the atmosphere through the natural process of photosynthesis and store the carbon (C) in their leaves, branches, stems, bark, and roots. The carbon sequestration benefit from plantation is determined by the difference in average carbon stock between the previous land use and the forest or plantation. Generally, predictions of the sequestration rate of different tree species cannot be made, since growth and sequestration depend on local climate, soil factors, and management. The rate of carbon sequestration depends on the growth characteristics of the tree species, the conditions for growth where the tree is planted, and the density of the tree's wood. It is the greatest in the younger stages of tree growth, between 20 and 50 years. Further complicating the issue is the fact that far less research has been done on tropical tree species as compared to temperate tree species.

The bhata land soil is dominated in Chhattisgarh and is categorized as waste land because it only supports pastures due to nutrient deficiencies and adverse physical properties. After tedious efforts, only few tree species survive in bhata land. In the present study, 25-year-old forest plantations were selected to assess the magnitude of growth, biomass, and carbon sequestration by different species with an objective to evaluate the potential tree species for bhata land in relation to changing climate.

Materials and Methods

Study area: The experiment site is located between 82.15°E longitude and 22.09°N latitude in Bilaspur, district of Chhattisgarh at an altitude of about 264 m above mean sea level. It is roughly 19 km south to the Bilaspur city and the soil of the site is red entisol with red oxide deposits and is nutrient deficient. The climate is pleasant and mild in the winter (minimum temperature 10 °C). There are medium rains (1320 mm) in the monsoon season. The summers are very hot and dry, with maximum temperature 45 °C. The selected area is only 10 km away from industrial zone of the Bilaspur city, where steel, motor, chemical and fertilizer industries, power plants, etc., are located.

Demarcation and enumeration of diameter and height: In order to understand how carbon dioxide sequestration patterns vary among plantation types, we estimated CO_2 sequestration in eight tropical plantations. The present study was carried out in Chakrabhatha Plantation Block of Bilaspur district during the year 2014. The whole plantation extent in over 100 ha using Kala siris (*Albizia lebbeck*), Neem (*Azadirachta indica*), Shisham (*Dalbergia sissoo*), Nilgiri (*Eucalyptus globulous*), Aonla (*Emblica officinalis*), Arjun (*Terminalia arjuna*), Peltaphorum (*Peltophorum ferruginium*), and Karanj (*Pongamia pinnata*) species in 4 × 4 m distance accommodating 625 plants per hectare.

After survey of the entire area, trees were enumerated according to diameter at breast height in 50 × 50 m sample plots. In total, 156 trees were considered from each species in order to determine the diameter at breast height (DBH) and height by using measuring tape and abney's level.

Estimation of carbon storage and CO_2 sequestration in tree: For each tree sampled, carbon storage was estimated using biomass and growth equation. To convert biomass to carbon storage, we assumed a constant carbon content of 50% for trees (Baishya et al. 2009). Estimations of biomass of tree were made as per formula $W = 0.15D^2H$ and the weight of carbon dioxide sequestered were calculated by multiplication of the weight of carbon in the tree by 3.66.

Soil organic carbon analysis: Soil samples were collected by dividing each main plot (50 × 50 m) area into three subareas. Representative soil samples from each subarea were collected by digging three pits of 30 cm wide, 30 cm deep, and 50 cm in length. Composite samples from all three subareas were obtained. Soil samples were air dried in shade, ground with wooden pestle, passed through a 2 mm sieve mesh, and stored in cloth bags for further analysis. Percent organic carbon was determined by Walkey and Black (1934) rapid titration method. In this method, 1.0 g soil was digested with a mixture of potassium dichromate (10 ml) and concentrated sulphuric acid (20 ml). The excess of potassium dichromate not reduced by the organic matter of the soil was determined by titration using standard ferrous ammonium sulphate solution in the presence of ortho-phosphoric acid using diphenylamine as an indicator.

Statistical analysis: The data were statistically analysed by the analysis of variance technique (ANOVA) using computer software SX.

Results and Discussion

Growth and Biomass of Different Tree Species

Field data of trees from study site were presented in Table 1, which reveal that the diameter and height attributes significantly varied among tree species. *A. lebbeck* and *Eucalyptus globulous* were the top most species with 43.14 and 29.94 cm diameter, respectively. Diameter of the other species was in between 14.27 and 19.68 cm. Similar pattern was also noticed in height of tree plantation, where both A. *lebbeck* and *E. globulous* achieved over 21 m height in 25 year of planting, while the same attribute in other tree species ranged between 13 and 15 m.

The biomass of different tree plantations is summarized in Table 2 which also demonstrates the same that.

A. lebbeck and *E. globulous* were accumulated 5.36 and 2.62 times more biomass, respectively, as compared to the average biomass of other six species

Table 1 Diameter and height status of different tree species after 25 years of planting in entisol soil

S. No.	Species	Plot size (m)	Density	Diameter (cm)	Height (m)
1	*Albizia lebbeck*	50 × 50	156	43.12	21.40
2	*Azadiractha india*	50 × 50	156	19.68	14.40
3	*Dalbergia sissoo*	50 × 50	156	15.67	14.40
4	*Emblica officinalis*	50 × 50	156	18.85	14.40
5	*Eucalyptus globulous*	50 × 50	156	29.94	21.60
6	*Peltophorum ferruginium*	50 × 50	156	18.09	14.20
7	*Pongamia pinnata*	50 × 50	156	14.27	13.40
8	*Terminalia arjuna*	50 × 50	156	18.15	15.00
	CD at $<p$ 0.05			4.50	1.05

Table 2 Biomass accumulation and carbon dioxide sequestration in forest tree plantations at 25 year age in entisol soil

S. No.	Species	Dry biomass (kg/tree)	CO_2 sequestered (Q/tree)
1	*Albizia lebbeck*	2625.13	21.82
2	*Azadiractha india*	616.96	5.13
3	*Dalbergia sissoo*	391.03	3.25
4	*Emblica officinalis*	566.14	4.70
5	*Eucalyptus globulous*	1284.63	10.68
6	*Peltophorum ferruginium*	513.93	4.27
7	*Pongamia pinnata*	301.70	2.50
8	*Terminalia arjuna*	546.71	4.54
	CD at $<p$ 0.05	27.29	0.12

plantations at the same age. The performance of *D. sissoo* and *P. pinnata* was observed as poor in bhata land as for as biomass accumulation was concerned.

CO_2 Sequestration of Different Tree Species

The data on CO_2 sequestration of tree species in plantation block of Chakarbhata, Bilaspur are depicted in Table 2. Results indicated that the CO_2 sequestration pattern varied greatly among forest plantations and was highly dependent on growth and biomass accumulation traits of the trees. On the basis of CO_2 sequestration potėntial of all the tree species under present investigation, tree species can be classified into three categories. *A. lebbeck* (21.82 q/tree) and *E. globulous* (10.68 q/tree) were in first category with highest degree of CO_2 sequestration, while in the second category *A. indica* (5.13 q/tree), *E. officinalis* (4.70 q/tree), *P. ferruginium* (4.27 q/tree) and *T. arjuna* (4.54 q/tree) were identified with medium level of CO_2 storage in their biomass. In third category, CO_2 sequestered tree species *D. sissoo* and *P. pinnata* were placed with the lowest CO_2 sequestration in bhata land soil. The variation in biomass accumulation and CO_2 sequestration rate by different tree species might be due to the variation in growth pattern exhibited by tree species in bhata land wasteland area. Therefore, the better performed tree species scored higher growth, biomass, and CO_2 sequestration in comparison to poor performer tree species, viz., *D. sissoo* and *P. pinnata*. Yin et al. (2012) also demonstrated that there was a difference in carbon content for the tree species due to variations in growth and increment pattern of the species. In addition, Bohre et al. (2012), Pandya et al. (2013) and Suryawanshi et al. (2014) also reported similar results in their investigation.

Soil Organic Carbon Under Different Tree Species

Soil organic carbon in rhizospheric soil of different tree plantations is an indicative of the positi impact of tree species in the improvement of soil. Generally, the tree plantations exhibit differential pattern of leaf fall depends upon the temperature and rainfall of the area, organic carbon accumulation, and varied under various tree plantations (Fig. 1). Soil organic carbon was found to enrich highest under *E. globulous* (0.52%) and than in *A. lebbeck* (0.51%) *E. officinalis* (0.51%) and *T. arjuna* (0.51%). However, the organic carbon accumulation was at par in *A. indica* (0.50%) and *P. pinnata* (0.50%) as compared to species recorded maximum SOC. The lowest carbon percent was recorded in *D. sissoo* (0.40%) and *P. ferruginium* (0.44%). This variation in SOC might be related to litter fall and size of canopy area. Moreover, Gupta and Pandey (2008) propounded that the variation in soil organic carbon pool was due to the litter fall pattern of the forest species and prevailing locality factors of the area as similar to our study.

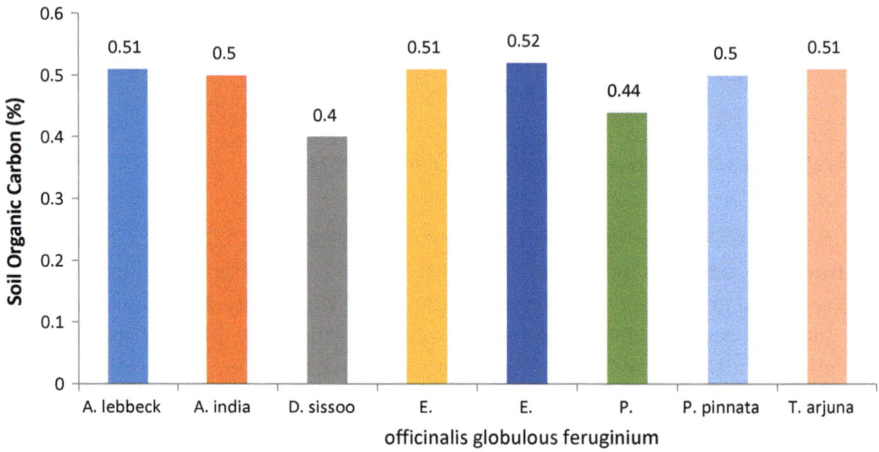

Fig. 1 Status of mean soil organic carbon under different tree plantations in entisol soil

Conclusion

Global warming is one of the most extensive concerns all over the world for the survival of earth and its biotic component in future. Under this circumstance, forest trees are playing their highest role in mitigating the adverse effect of global warming through CO_2 absorption, water body recycling, adding of organic matter, and maintain microbial populations for soil fertility. As trees vary in growth and biomass status from locality to locality, it is pertinent to evaluate the potential of different tree species plantations on different attributes for various types of soil, zones, and locality. In present investigations, *E. globulous* was found to be most effective in entisol soil as it rendered the maximum height while the breast height diameter was found the maximum in *A. lebbeck.* The growth performance of *P. pinnata* was recorded minimum. It is concluded that the CO_2 sequestration rate is dependent on growth status of the tree species. The more the biomass means the higher the CO_2 sequestration. *A. lebbeck* and *E. globulous* accumulated higher biomass compared to other species in entisol soil and thus were more efficient in carbon shrink. Moreover, the soil under these species consisted higher percentage of organic carbon which again proves their efficiencies for carbon shrink and may be recommended for such types of waste land sites.

References

Baishya R, Barik SK, Upadhaya K (2009) Distribution pattern of aboveground biomass in natural and plantation forests of humid tropics in northeast India. Trop Ecol 50:295–304

Bohre P, Chaubey OP, Singhal PK (2012) Biomass accumulation and carbon sequestration in Dalbergia sissoo. Roxb. Int J Biosci Biotechnol 4(3):29–44

Dhruw SK, Singh LJ, singh AK (2009) Storage and sequestration of carbon by leguminous and non leguminous trees on red lateritic soil of Chhattisgarh. Ind For 135(4):531–538

Gupta MK, Pandey R (2008) Soil organic carbon pool under different plantations in some district of Uttarakhand and Haryana. Ind J For 31(3):369–374

IPCC (2003) Good practice guidance for land use, land use change and forestry. Published by the institute for global environmental strategies for the IPCC. Publishers Institute for global environmental strategies, Japan

Pandya IY, Salvi H, chahar O, Vaghela N (2013) Quantitative analysis on carbon storage of 25 valuable tree species of Gujarat, Incredible India. Ind J Sci Res 4(1):137–141

Suryawanshi MN, Patel AR, Kale TS, Patil PR (2014) Carbon sequestration potential of tree species in the environment of North Maharastra University campus, Jalgaon [MS] India. Biosci Dis 5(2):175–179

Walkey A, Black CA (1934) An examination of the method for determining soil organic matter and a proposed modification of the chromic acid titration method. Soil Sci 37(16):29–39

Yin W, Yin M, Zhao L, Yand L (2012) Research on the measurement of carbon storage in plantation tree trunks based on the carbon storage dynamic analysis method. Int J For Res 2012:1–10

Impact of Copper Oxide Nanoparticles on Growth of Different Bacterial Species

Tapan Adhikari, Garima Dube, S. Kundu and A.K. Patra

Abstract Copper oxide nanoparticles have numerous exciting applications due to their unique chemical and physical properties. With increased applications of engineered nanostructures, releases of such materials to soil are undoubtedly inevitable. Their potential environmental risks have attracted increasing concern. One area of great concern is their effect on microorganisms, which are important components of ecosystems. Impact of copper oxide nanoparticles (<50 nm) was assessed by using plate assay method on *Bacillus subtilis and Escherichia coli* using nutrient agar media. A parallel experiment was also conducted with its ionic form through $CuSO_4 \cdot 5H_2O$. *B. subtilis* and *E. coli* showed no growth beyond concentration of 20 mg Cu/kg of normal copper ion using $CuSO_4 \cdot 5H_2O$. While in case of nanocopper through CuO (<50 nm), *B. subtilis* recorded no growth above 10 mg Cu/kg and *E. coli* with least susceptibility showing no growth above 40 mg Cu/kg. Experimental results revealed that CuO (<50 nm) affected the Gram-positive and Gram-negative bacteria differently. These findings are of great help toward building a comprehensive understanding of the potential impact of nanoparticles on the environment.

Keywords Copper oxide nanoparticle · *Bacillus subtilis* · *Escherichia coli*
Nutrient agar media

Introduction

Amongst the various transition metal oxides, copper oxide (CuO) has attracted greater attention due to its fascinating properties such as the basis of high critical temperature superconductors. CuO nanoparticle is a semiconducting compound with a narrow band gap and is used for photoconductive and photo-thermal applications. It has wide variety of applications, including use in medical diagnostics,

T. Adhikari (✉) · G. Dube · S. Kundu · A.K. Patra
Indian Institute of Soil Science, Nabibagh, Berasia Road, Bhopal 462038, India
e-mail: tapan_12000@rediffmail.com

V.P. Singh et al. (eds.), *Energy and Environment*, Water Science and Technology Library 80, https://doi.org/10.1007/978-981-10-5798-4_5

controlled drug release, separation technologies, and environmental engineering. The bactericidal property of such nanoparticle depends on their size, stability, and concentration added to the growth medium, since this provides greater retention time for bacterium–nanoparticles interaction. In general, bacterial cells are in the micron-sized range. Most bacterial cells have cellular membranes that contain pores in the nanometer range. A unique property of crossing the cell membrane can potentially be attributed to synthesized nanoparticles through such bacterial pores. However, to make this possible, it is important to overcome challenges and to prepare/design nanoparticles which are stable enough to significantly restrict bacterial growth while crossing the cell membrane. As manufacturers increase their nanomaterial production to meet ever-increasing demands, the release of such materials into soil is inevitable (Lee et al. 2010). Increasing concerns have been raised on how this release would affect ecosystem health and human safety (Klaine et al. 2008). A few studies have been performed to elucidate the mechanism of bactericidal action of nanoparticles. The other researcher (Tsao et al. 2002) suggested that the exposure of Gram-positive bacteria to carboxyfullerene nanoparticles resulted in the puncturing of the bacteria leading to cell death. Another proposed way in which the membrane can be compromised is the alteration of membrane lipid components (Koch et al. 2005). Unfortunately, little knowledge is available to date despite these concerns. A paramount aspect of understanding this impact is to determine how microorganisms respond to copper dioxide nanoparticle because these organisms are an indispensable part of the environment. Microorganisms are the drivers of global biogeochemical cycles. They are involved in the cycling of carbon, nitrogen, sulfur, and phosphorus. Because microorganisms are especially sensitive to environmental changes (Stadtman 1990), the structure and abundance of the microorganism community may shift in response to foreign nanomaterials (Ge et al. 2011). Because microorganisms help to regulate and maintain the overall ecosystem's health and function, changes in the microbial community will also have a great effect on the entire ecosystem (Kanerva et al. 2008). Therefore, a better understanding of how microorganisms respond to nanomaterials can help to address environmental and health concerns brought about by the manufacture and use of nanomaterials. Against this backdrop, an attempt was made to investigate the effect of CuO nanoparticles (NPs) (<50 nm) on the growth of different bacterial species.

Materials and Methods

Gram-positive bacteria *Bacillus subtilis* and Gram-negative bacteria *Escherichia coli* were cultured on nutrient agar plates and then tested against different concentration of normal copper ion from $CuSO_4{\cdot}5H_2O$ and copper nanoform from CuO (<50 nm) (0, 2, 4, 6, 8, 10, 20, 40, 60, 80, and 100 mg/L) on nutrient agar plates prepared by using Nutrient Agar Media (Hi Media, India). The growth was determined by plate assay in which first of all dilution series of bacterial isolates were prepared and then from the fifth dilution culture was inoculated on nutrient agar plates of various copper concentration in both the forms through spread plate method and then incubated at 30 °C for 24 h. The colonies grown were then counted.

The CuO nanoparticles (NPs) were purchased from Sigma-Aldrich Company, St. Louis, MO, USA with a purity of 99.5%, particle size of <50 nm and a surface area of 45 ± 10 m^2/g. The morphology of the CuO nanoparticles was examined using transmission electron microscopy (TEM, JEOL, 100 CX, Japan).

Results and Discussion

The TEM image of the copper oxide NPs revealed their spherical, truncated, and uneven nature with an average size of approximately 50 ± 20 nm (Fig. 1a). The TEM micrographs indicated that the copper oxide nanoparticles were mono dispersed with a narrow size distribution and near spherical morphology. Analysis of particles in TEM monograph indicated hexagonal particles with the average size of 50 nm. In the tangential interaction of copper oxide NPs, the existence of a lateral force causes the aggregation of the NPs. The zeta potential of CuO nanoparticles was −1.91 mv (Fig. 1b). Antimicrobial action of copper in both ionic and nanoparticle form varied with each bacterial isolate. *B. subtilis*, a Gram-positive bacteria having the ability to form a tough endospore, allowing the organism to tolerate extreme environmental conditions, when grown against normal copper ion showed tolerance level at 20 mg Cu/kg while when grown against copper in nanoparticle form showed tolerance level at 10 mg Cu/kg (Table 1). Gram-negative bacteria, *E. coli* showed tolerance level at 20 mg Cu/kg when grown against normal copper ion doses, however, when grown against copper in nanoparticle form showed tolerance level at 40 mg/kg. In this study, the copper oxide nanoparticles showed remarkable antibacterial activity against both Gram-positive (*B. subtilis*) (Plate 1) and Gram-negative (*E. coli*) bacteria (Plate 2). Growth inhibition occurs already after 1-day metal treatment (Weckx and Clijsters 1996). The mode of action of copper ions was shown by its action on enzymes. Under laboratory conditions, copper have been shown to eliminate up to 99% of germs in the shortest period of time. Copper ions affect bacteria in two sequential steps (a) direct interaction between the surface and the bacterial outer membrane (b) second is related to holes in the outer membrane, through which the cell loses vital nutrients and water, causing a general weakening of cells (De Vos et al. 1989; De Vos et al. 1991). It is strongly suspected that when a bacterium comes in contact with Cu surface, a short circuiting of current in cell membrane can occur. This weakens the membrane and creates hole. Another way to make holes in membrane is by localized oxidation or 'rusting' which is a hit occurs in presence of oxygen.

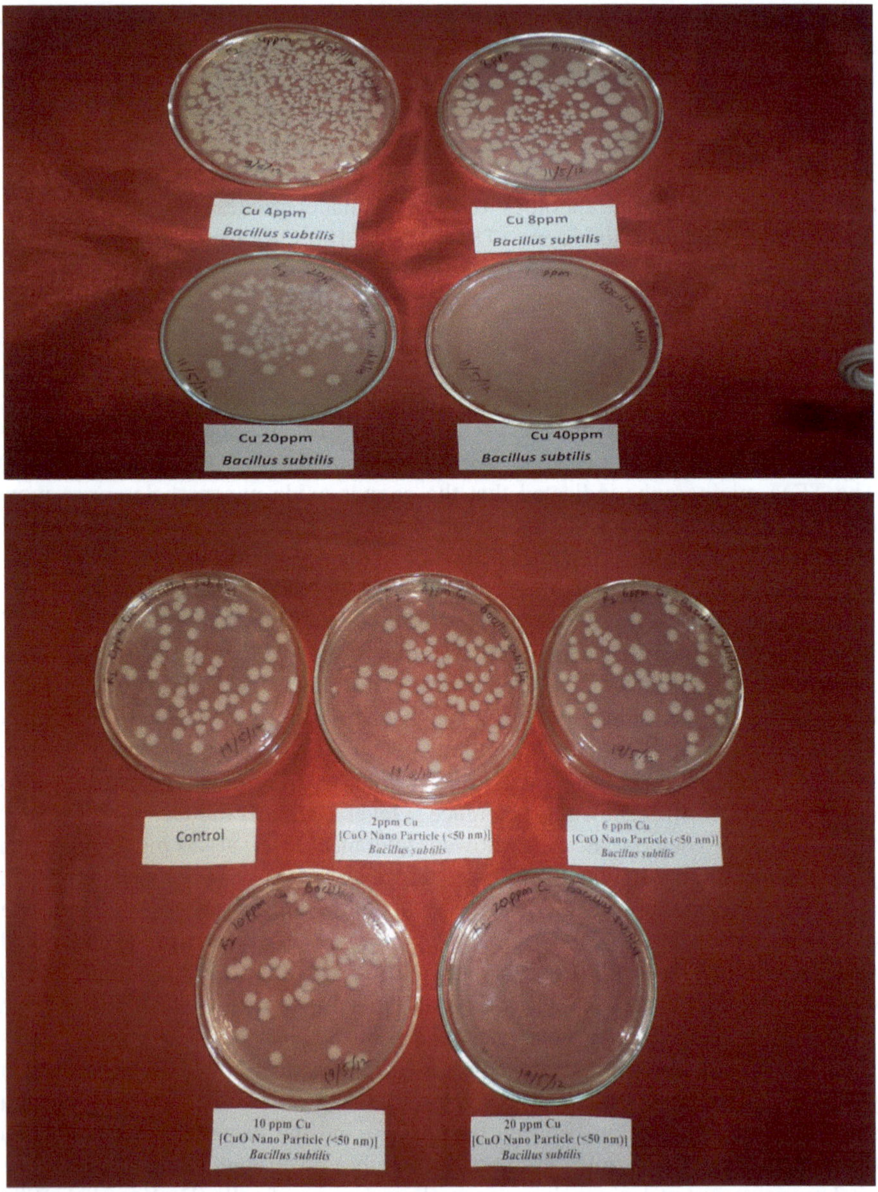

Plate 1 Effect of copper in both ionic and nanoform on growth of *Bacillus subtilis*

Now the cell's main defense (outer envelop) has been breached, there is an unopposed stream of copper ions entering the cell which put several vital processes running in cells into danger. The reactions are accomplished and catalyzed by

Table 1 Effect of copper in both ionic and nanoform on growth of different bacteria

S. No.	Copper doses (mg/kg)	*Bacillus subtilis* ($\times 10^7$) (CFU/ml)		*Escherichia coli* ($\times 10^7$) (CFU/ml)	
		Cu^{+2}	Cu (nano)	Cu^{+2}	Cu (nano)
1	0	100 ± 1.73	104 ± 1.73	100 ± 2.03	100 ± 2.03
2	2	96 ± 1.15	45 ± 0.58	16 ± 0.58	45 ± 0.67
3	4	88 ± 1.15	10 ± 0.58	15 ± 0.33	40 ± 0.65
4	6	46 ± 0.88	6 ± 0.48	14 ± 0.45	38 ± 0.33
5	8	26 ± 0.80	4 ± 0.55	12 ± 0.33	34 ± 0.35
6	10	14 ± 0.58	3 ± 0.45	8 ± 0.39	29 ± 0.55
7	20	12 ± 0.77	tr	7 ± 0.35	26 ± 0.32
8	40	tr	tr	tr	6 ± 0.43
9	60	tr	tr	tr	tr
10	80	tr	tr	tr	tr
11	100	tr	tr	tr	tr

Sample Number 3, Mean ± Standard Error, *tr* trace

enzymes. When excess copper binds to these enzymes, their activity grinds to halt. Also excess of copper can induce a number of free radical processes in protein (Stadtman 1990) and lipid cell membrane components (Strange and Macnair 1991) causing destabilization of membranes and increase in their permeability (Wainwright and Woolhouse 1977). Copper oxide nanoparticle has been characterized, both physically and chemically and investigated with respect to potential antimicrobial applications. It remains unclear as to the precise mechanism by which copper nanoparticles exert their antimicrobial activity. The experimental results revealed that the antimicrobial effect of nanocopper is more than that of nanocopper ion form in case of Gram-positive bacterial species, i.e., *B. subtilis* while in case of Gram-negative bacteria *E. coli,* normal copper ion antimicrobial activity seems to be more effective as compared to copper in nanoparticle form. This difference in antimicrobial effectiveness between Gram-positive and Gram-negative bacteria may be due to the difference in the thickness of peptidoglycan layer present in the membrane of bacteria. Copper in its ionic form is a required trace element for most pro- and eukaryotic organisms, including humans. Though needed in small amounts, copper can easily become toxic when in surplus. This toxicity is caused mainly by the intrinsic properties of copper, as free copper ions undergo redox cycling reactions alternating between Cu(I) and Cu(II). This also results in the transfer of electrons to hydrogen peroxide and the concomitant generation of hydroxyl radicals that readily attack and damage cellular biomolecules (Maksymiec and Baszynski 1996a, b). Recently, it was found that the majority of copper stress in *Escherichia coli*, as indicated by hydroxyl radical formation, occurs within the periplasm, away from the cytoplasmic DNA (Macomber et al 2007). The cytoplasm might thus be better protected from copper-mediated oxidative stress, and indeed cells usually prevent accumulation of significant intracellular concentrations of free copper ions either by producing copper binding chaperones (Magnani et al. 2008)

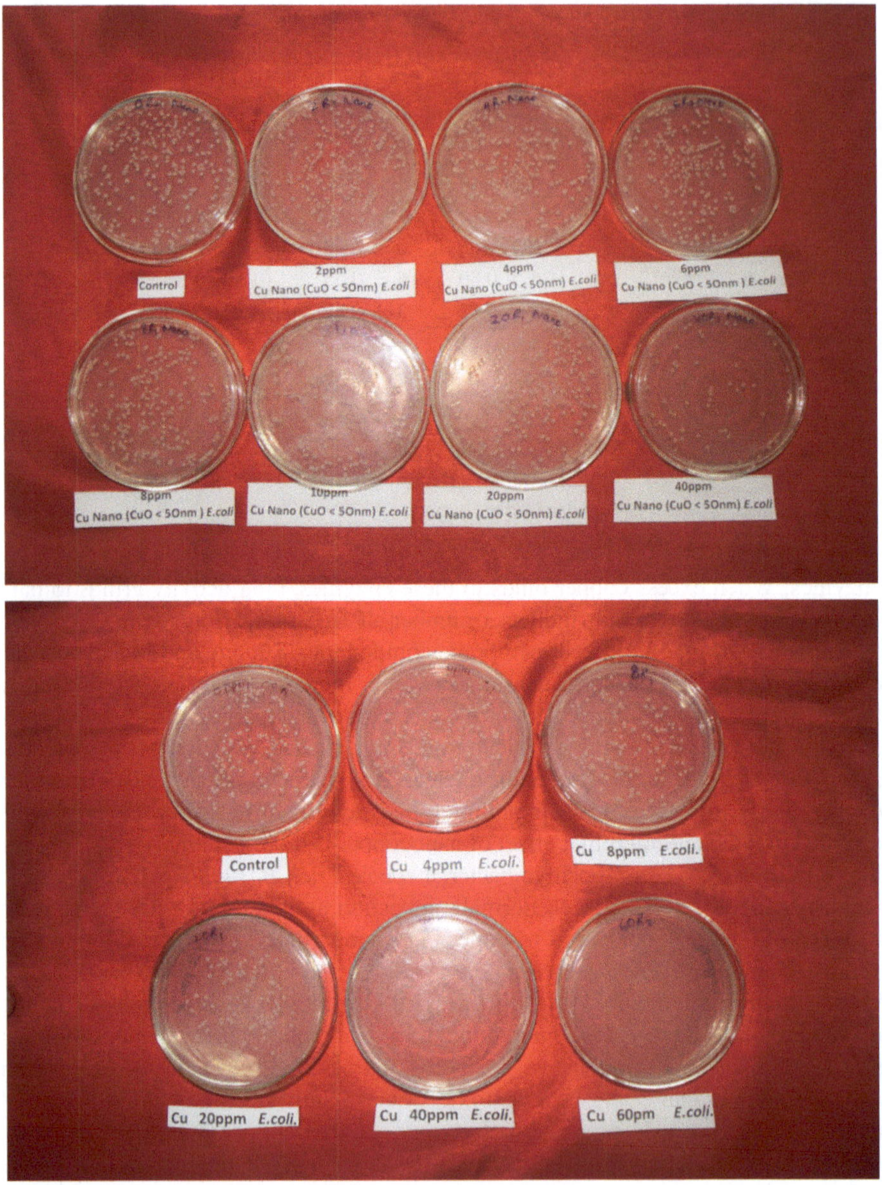

Plate 2 Effect of copper in both ionic and nanoform on growth of *Escherichia coli*

or unspecific chelators such as glutathione (Miras et al. 2008) or by efflux (Rensing et al. 2000). Nevertheless, copper ions within the cytoplasm also cause damage. Surprisingly, this damage is not related to oxidative stress but is exerted directly by

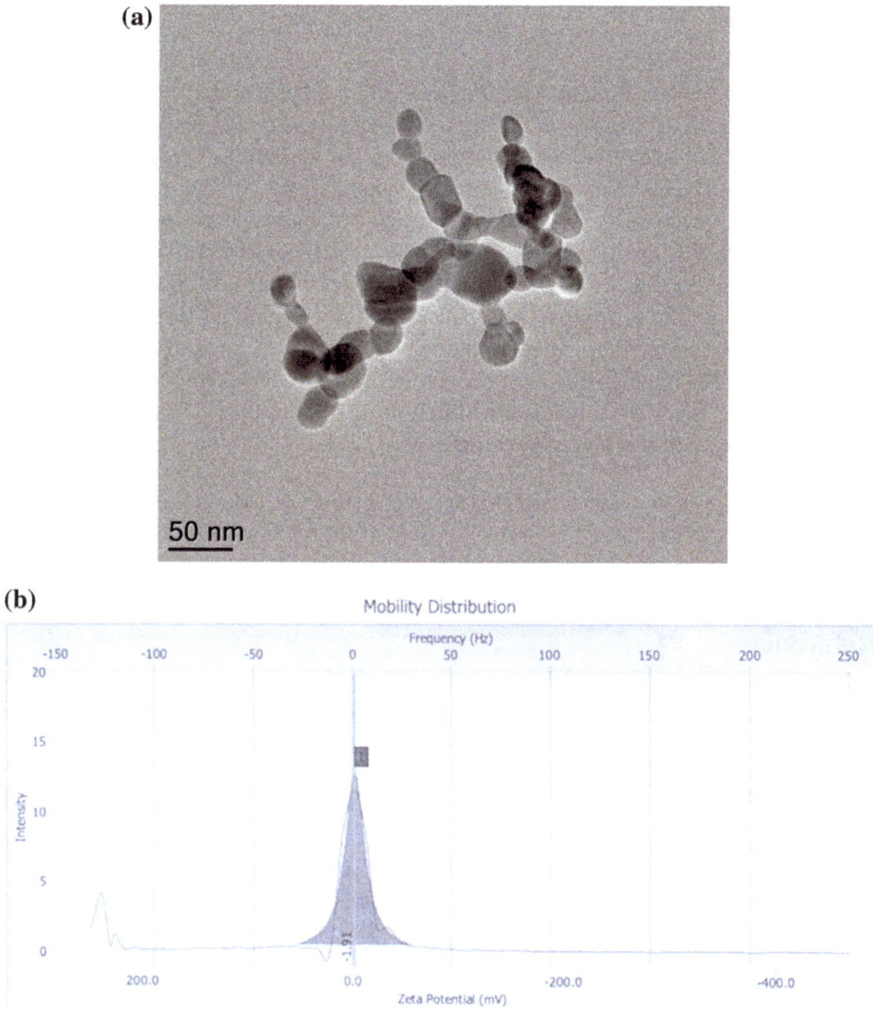

Fig. 1 **a** Transmission electron microscope photograph of copper nanoparticle **b** Zeta potential of CuO nanoparticles

the metal ions. It seems that copper ions attack and displace iron atoms from enzymes with solvent-exposed iron sulfur clusters such as those of hydratases (Macomber and Imlay 2009). Thus, the presence of oxygen is not needed for this reaction, and there is no copper mediate oxidative stress involved in this damage (Macomber and Imlay 2009).

Conclusions

It is clear from the experimental results that CuO nanoparticles have shown greater antimicrobial activity against *B. subtilis.* The variation in the sensitivity or resistance to both Gram-positive and-negative bacteria populations could be due to the differences in the cell structure, physiology, metabolism, or degree of contact of organisms with nanoparticles. Greater sensitivity among Gram-positive bacteria such as *B. subtilis* to the CuO nanoparticles has been attributed to the greater abundance of amines and carboxyl groups on their cell surface and greater affinity of copper toward these groups. Alternatively, Gram-negative bacteria like *E. coli* have a special cell membrane structure which possesses an important ability to resist antimicrobial agents. Furthermore, other factors such as nanoparticle diffusion rates may also affect bacterial strain differently. Nevertheless, further studies are required to illustrate the mechanism properly.

Acknowledgements We sincerely acknowledge World Bank and Indian Council of Agriculture Research (ICAR) for the financial support under National Agricultural Innovative Project (NAIP).

References

De Vos CHR, Schat H, Vooijs R, Ernst WHO (1989) Copper induced damage to the permeability barriers in roots of *Silene cucubalus.* J Plant Physiol 135:164–169

De Vos CHR, Waal MAM, Vooijs R, Ernst WHO (1991) Increased resistance to copper induced damage of root cell plasma lemma in copper tolerant *Silene cucubalis.* Physiol Plant 82:523–528

Ge Y, Schimel JP, Holden PA (2011) Evidence for negative effects of TiO_2 and ZnO nanoparticles on soil bacterial communities. Environ Sci Technol 45:1659–1664

Kanerva T, Palojarvi A, Ramo K, Manninen S (2008) Changes in soilmicrobial community structure under elevated tropospheric O_3 and CO_2. Soil Biol Biochem 40:2502–2510

Klaine SJ, Alvarez PJJ, Batley GE, Fernandes TF, Handy RD, Lyon DY, Mahendra S, McLaughlin MJ, Lead JR (2008) Nanomaterialsin the environment: behavior, fate, bioavailability, and effects. Environ Toxicol Chem 27:1825–1851

Koch AM, Reynolds F, Merkle HP, Weissleder R, Josephson L (2005) Transport of surface-modified nanoparticles through cellmonolayers. Chem Bio Chem 6:337–345

Lee J, Mahendra S, Alvarez PJJ (2010) Nanomaterials in the construction industry: a review of their applications and environmental health and safety considerations. ACS Nano 4:3580–3590

Macomber L, Imlay JA (2009) The iron-sulfur clusters of dehydratases are primary intracellular targets of copper toxicity. Proc Natl Acad Sci USA 106:8344–8349

Macomber L, Rensing C, Imlay JA (2007) Intracellular copper does not catalyze the formation of oxidative DNA damage in *Escherichia coli.* J Bacteriol 189:1616–1626

Magnani D, Barre O, Gerber SD, Solioz M (2008) Characterizationof the CopR regulon of *Lactococcus lactis* IL1403. J Bacteriol 190:536–545

Maksymiec W, Baszynski T (1996a) Chlorophyll fluorescence in primary leaves of excess Cu treated runner bean plants depends on their growth stages and the duration of Cu action. J Plant Physiol 149:196–200

Maksymiec W, Baszynski T (1996b) Different susceptibility of runner bean plants to excess copper as a function of growth stages of primary leaves. J Plant Physiol 149:217–221

Miras R, Morin I, Jacquin O, Cuillel M, Guillain F, Mintz E (2008) Interplay between glutathione, Atx1 and copper. 1. Copper(I) glutathionate induced dimerization of Atx1. J Biol Inorg Chem 13:195–205

Rensing C, Fan B, Sharma R, Mitra B, Rosen BP (2000) CopA: an *Escherichia coli* Cu(I)-translocating P-type AT pase. Proc Natl Acad Sci USA 97:652–656

Stadtman ER (1990) Metal ion-catalyzed oxidation of proteins: biochemical mechanism and biological consequences. Free Radical Biol Med 9:315–325

Strange J, Macnair MR (1991) Evidence for a role for the cell membrane in copper tolerance of *Mimulus guttatus* fisher ex DC. New Phytol 119:383–388

Tsao N, Luh TY, Chou CK (2002) In vitro action of carboxyfullerene. J Antimicrob Chemother 49:641–649

Wainwright SJ, Woolhouse HW (1977) Some physiological aspects of copper and zinc tolerance in *Agrostis tenuis* Sibth: cell elongation and membrane damage. J Exp Bot 28:1029–1036

Weckx JEJ, Clijsters HMM (1996) Oxidative damage and defense mechanisms in primary leaves of *Phseolus vulgaris* as a result of root assimilation of toxic amount of copper. Physiol Plant 96:506–512

Performance of Low-Cost Microbial Fuel Cell Using Earthenware Separator

Sudhansu Behera and Manaswini Behera

Abstract The extensive use of fossil fuels and the associated problems of environmental pollution and global warming demand alternate renewable energy sources. Microbial fuel cells (MFCs) is an emerging technology which uses the chemical energy stored in the organic matter and generates society's most useful form of energy, i.e., electricity. MFC has the advantages of less sludge generation, low temperature operation, and omission of off-gas treatment. The performance of a dual-chambered low-cost MFC employing earthenware separator modified with montmorillonite was studied as a potential device for dairy wastewater treatment and simultaneous bioelectricity generation. The dual-chambered MFC with inner anodic chamber and concentric outer cathode chamber showed appreciable performance with a maximum power density of 33.44 mW/m^2 normalized to anodic electrode surface area while fed with synthetic dairy wastewater having chemical oxygen demand (COD) of 1920 $\pm$ 20 mg/L. The wastewater was treated anaerobically in the anode chamber and the effluent from the anode chamber was given aerobic treatment in the cathode chamber. The MFC has shown a maximum COD removal efficiency of 89%. The low-cost MFC showed promising performance, which can be employed as a treatment technology for organic wastewater simultaneously generating electricity.

Keywords Microbial fuel cell (MFC) · Earthenware separator
Chemical oxygen demand (COD) · Power density
Proton exchange membrane (PEM)

Introduction

Fossil fuel depletion and energy crisis drive attention worldwide to renewable energy sources. The microbial fuel cell (MFC) is a representative of new biocatalyzing technology that can directly produce electricity from oxidation of organic

S. Behera (✉) · M. Behera
School of Infrastructure, Indian Institute of Technology, Bhubaneswar 751013, India
e-mail: sb33@iitbbs.ac.in

V.P. Singh et al. (eds.), *Energy and Environment*, Water Science
and Technology Library 80, https://doi.org/10.1007/978-981-10-5798-4_6

matters using bacteria (Chaudhuri and Lovely 2003). The transformation of chemical energy stored in the bonds of organic matter to useful electrical energy is governed by the microorganisms present in wastewater, which act as active bio-catalyst in anaerobic condition. The direct electricity thus produced is devoid of pollutants as the net carbon emission in the process is zero. The simultaneous treatment of wastewater and bioenergy production in a noble device MFC overcomes many challenges in the development of such needs. The major challenges related to MFC are its high operational cost and potential upscaling of power output. Although extensive research is taking place for the development of such a low-cost noble MFC still there are many loop holes behind in the field of MFC materials.

Operating Principle of MFC

The MFC is operated by anaerobic degradation (oxidation) of organic matter by the anaerobes in anode chamber consequently producing electrons that travel through a series of respiratory enzymes in the cell and provide energy in the form of ATP (adenosine triphosphate). The electron transfer mechanism by the bacteria is so far known to occur through two ways: (i) by electron shuttle via self-produced mediator (Rabaey et al. 2004) and (ii) nanowires produced by both *Geobacter* and *Shewanella* species (Gorby and Beveridge 2005). Thus, the electrons transfer to cathode through external electric circuit and are accepted at cathode chamber by terminal electron acceptors (e.g., oxygen, nitrate, sulfate, and others). The protons pass through cation exchange membrane to cathode which further react with electron and oxygen forming water as a by-product. The schematic of a traditional MFC is represented in Fig. 1.

The present study focused on fabrication of a low-cost dual-chambered MFC reactor by introducing a ceramic-based composite membrane. The earthenware separator was studied with a thickness of 5 mm for a period of 45 days. The anodic substrate as synthetic dairy wastewater was continuously supplied at flow rate of 2.16 mL/min followed by continuous aeration at cathode. The MFC was studied without any addition of external mediator in the electron transfer mechanism.

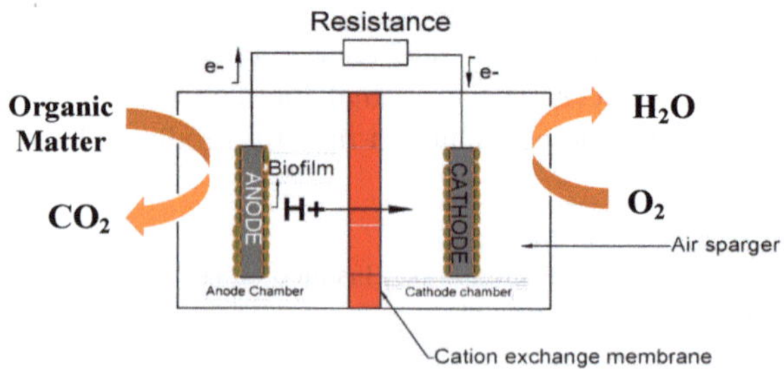

Fig. 1 A schematic of dual-chambered MFC

Methodology

MFC Construction

The dual-chambered MFC (Fig. 2) was fabricated using transparent poly acrylic material having inner cubical anode chamber inside a concentric outer cathode chamber. The anode chamber has a working volume of 780 mL while that of cathode is 6.6 L. Four grooves each having size of 6 cm × 8 cm were made for placing earthenware separator. Wastewater was fed to the anode chamber from the bottom. Two openings were made at the top of two opposite side walls of anode chamber for the passage of effluent from anode chamber to cathode chamber. Four earthenware separators having effective thickness of 5 mm were fixed on four sides of the anode chamber. Stainless steel (SS) mesh was used as anode electrode with total projected surface area 502.4 cm^2. Similarly graphite plates of size 6 cm × 8 cm having thickness of 0.5 cm were used as cathode electrode. The anode and cathode electrodes were placed very close to earthenware separator at a distance of 0.7 cm apart. Thin copper wire with full insulation was used in the electric circuit for conveyance of electricity. The anode chamber was kept airtight with a plate of poly acrylic material with stainless steel screws for tightening to maintain anaerobic condition.

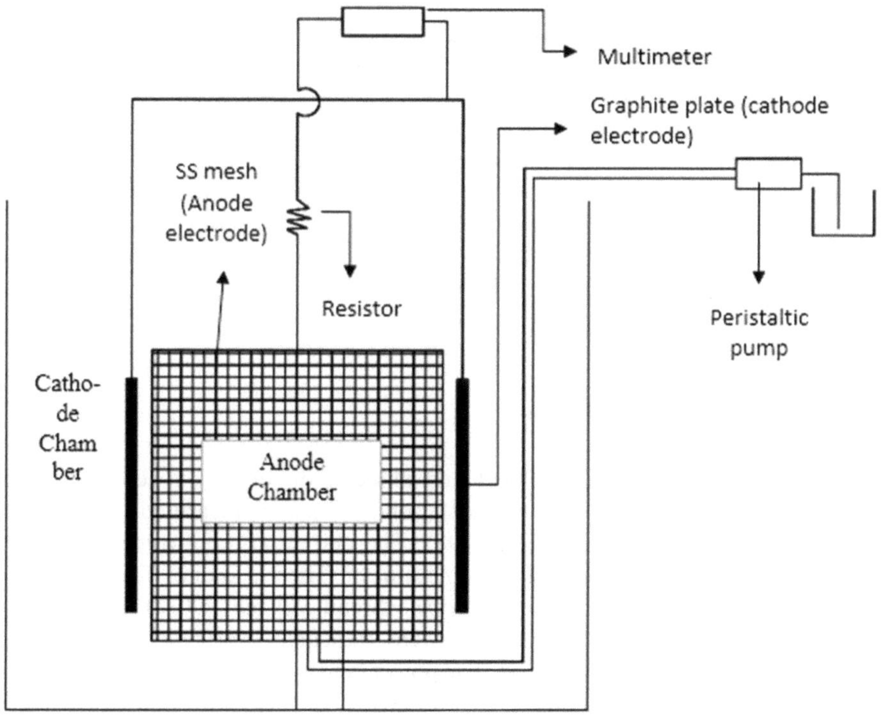

Fig. 2 A graphical presentation of the experimental setup of MFC

MFC Operation

The dual-chambered MFC was operated with a continuous supply of synthetic dairy wastewater having chemical oxygen demand (COD) of 1920 ± 20 mg/L. The substrate was supplied at a flow rate of 2.1 mL/min (hydraulic retention time, HRT = 6 h) by using a peristaltic pump. The constituents of synthetic dairy wastewater are milk powder, 1.6 g/L; $FeCl_3.6H_2O$, 15 mg/L; $MgSO_4.7H_2O$, 70 mg/L; $CaCl_2.H_2O$, 30 mg/L; $MnSO_4.7H_2O$, 15 mg/L; KH_2PO_4, 60 mg/L; NH_4Cl, 120 mg/L; $NaHCO_3$, 1 g/L (Raj and Murthy 1999; Zielinska et al. 2013). The substrate was given anaerobic treatment in the anode chamber then the effluent of anode chamber was given aerobic treatment in the cathode chamber with the help of active aeration. The anode chamber was inoculated with the bottom sludge of domestic sewer having volatile suspended solid (VSS) of 32.5 g/L. The anode and cathode electrodes were connected externally through insulated copper wire with a load of 150 Ω.

Preparation of Earthenware Separator

The earthenware separator was prepared by blending 80% of red soil (Laterite origin) and 20% of montmorillonite (cation exchanger) with the addition of small amount of water. Wet clay after achieving certain plasticity was filled in mould of size 7 cm × 9 cm × 0.5 cm. They are then exposed to open air for 5 days and dried in hot air oven at 150 °C for 2 h, and finally kept in muffle furnace at 650 °C for 30 min. Energy dispersive spectroscopy (EDS) analysis was carried out to find out the composition of earthenware separator. The composite membrane contained SiO_2—49.23%, Al_2O_3—21.6%, Fe—21.5%, K—2.15%, MgO—1.11%, Ti—1.45%, Na—0.95%, and Cu—2%. The photograph of clay mineral modified earthenware separator is shown in the following Fig. 3.

Analysis and Calculation

The COD concentrations of influent and effluent, dissolved oxygen (DO), pH, volatile suspended solid (VSS), and total solid (TS) were measured according to APHA standard procedure (APHA 1998). The current (I) and voltage (V) were measured by a true RMS professional multimeter (Extech, India). The power can thus be calculated as $P = V * I$. The open circuit voltage (OCV) and working voltage (external resistance of 150 Ω) were measured every day throughout a period of 45 days of working period. The power density and volumetric power were calculated normalized to anodic electrode area and net liquid volume of anode chamber, respectively. Polarization study was conducted after achieving a stable

Fig. 3 Fabricated earthenware proton exchange membrane

performance by varying external resistances from 50 to 4500 Ω. The internal resistance associated with MFC was calculated from the slope of line plotted between current and voltage.

The Coulombic efficiency (CE) is defined as the ratio of total coulombs recovered during the operation per total coulombs present in the substrate (Logan 2008). The CE of MFC operated in continuous mode of substrate supply can be found out from the formula:

$$\mathrm{CE} = \frac{8I}{Fq\Delta\mathrm{COD}}, \quad (1)$$

where, I is the measured electric current; 8 is a constant used for substrate measured in terms of COD based on molecular weight of O_2 as 32 g/mol; F is the Faraday's constant = 96485 C/mol; q is the flow rate of substrate in L/day; ΔCOD is the change in COD value of the influent and effluent in mg/L. The morphological characteristics of earthenware separator were studied by energy dispersive spectroscopy (EDS) analysis.

Results and Discussion

Physical Characterization of Separator

EDS Analysis

The microscopic analysis of earthenware PEM has shown an appreciable amount of silica and aluminum as the key inorganic elements in a typical soil sample. The chemical constituents of montmorillonite and the composite soil membrane are listed in Table 1.

The interpretation for the above elements present in the composite membrane shows abundant amount of oxides of silica which shows a potential membrane for proton exchange. The clay particles contain huge amount of silicon ion creating negative charge when combined with silicon–oxygen tetrahedron. Thus, the presence of silicon ions helps in retaining the positive ions, i.e., protons (H^+) ions on the surface of earthenware separator (Behera and Ghangrekar 2011).

The clay mineral montmorillonite is found to have highest cation exchange capacity (CEC) ranging about 70–100 meq/100 g among all minerals present in soil. The CEC is the number of exchangeable cations per dry weight that a soil is able to hold. Therefore, Blending low-cost and easily available cation exchangers could possibly enhance the performance. It was decided to blend different proportions of montmorillonite by weight with the parent material, i.e., red soil. Further, the clay minerals are very fine and by blending with soil having higher particle size could possibly reduce the voids. Thus, montmorillonite has the potential to acts as proton exchanger and plays a great role in the fabrication of earthenware separator.

Table 1 Composition of soil with their weight percentages

Elements	Montmorillonite (%)	Composite (Montmorillonite + red soil) (%)
Al_2O_3	19.68	25.64
SiO_2	51.42	45.78
K (Feldspar)	1.89	2.2
Fe	17.76	22.94
Cu	2.44	3.43
Ti	2.81	–
Na (Albite)	1.19	–
MgO	1.95	–
Ca (Wollastonite)	0.86	–

Table 2 Power production values

Working voltage (mV)	Power density (mW/m^2)	Volumetric power (W/m^3)	OCV (mV)
470 (max)	29.31	1.88	490
422 (average)	23.71	1.52	440

Power Production in MFC

The direct electricity production in MFC from organic substrates is a function of strength (COD) of organic matter. The more the rate of oxidation in anode chamber, the more would be the power density. A maximum open circuit voltage (OCV) of 0.49 V obtained on the 16th day of operation. The maximum and average power production values are shown in Table 2. The power was calculated by the formula $P = V * I$, where P is the power generated in watt, V is the working voltage and I is the measured current with a load of 150 Ω.

The volumetric power in the current study using a low-cost earthenware separator was found to be higher than the value 1.6 W/m^3 reported by Jana et al. (2010), used Nafion as potential PEM.

Coulombic Efficiency

The average CE during the period of stable performance was found to be 4.16%. The low CE shown by the MFCs might be due to utilization of large percentage of substrate by anaerobic microbes such as methanogenic archaea or sulfate-reducing bacteria other than anode-respiring bacteria (He et al. 2005).

Polarization Study and Internal Resistance

The polarization study is mainly conducted to know the effect of different external resistances on the performance of the system. Figure 4 shows the polarization curve as a function of current density, voltage and power density measured at variable external circuit load (50–4500 Ω). During polarization, current generation decreased with increase in external resistance, which indicated typical fuel cell behavior. The power density curve showed a maximum power density of 33.44 mW/m^2 at an external resistance of 200 Ω during polarization.

From the curve between current (I) and voltage (V), the slope of line represents internal resistance of the reactor which was found to be 108 Ω.

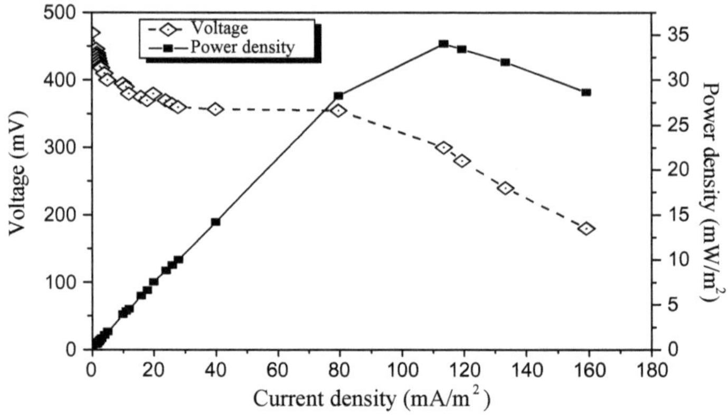

Fig. 4 Voltage and power density values as function of current density

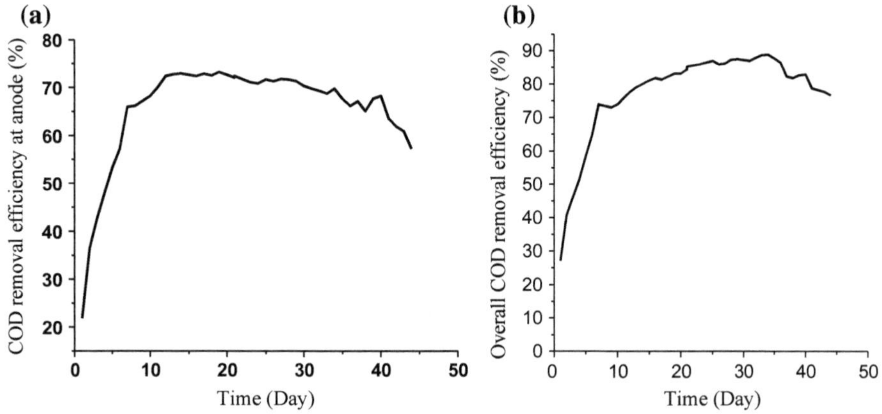

Fig. 5 **a** COD removal in anode chamber. **b** Overall COD removal

Wastewater Treatment

There has been a consistent increase in the COD removal efficiency after an acclimatization period of 12 days of operation. The average COD removal efficiency of anode chamber after attaining a stable removal rate was 69.7%. Similarly, the COD removal efficiency of cathode chamber was calculated to be 45.2% and finally the overall COD removal efficiency was calculated to be 83.3%. The COD removal efficiency decreased after 35th day of operation, which may be attributed to the decrease in the room temperature in the winter. The dual-chambered MFC was found to be efficient to treat synthetic dairy wastewater. The introduction of biocathode also helped in reducing the strength of organic matter of anode effluent by continuous aeration. Figure 5a, b shows COD removal at anode chamber and the overall COD removal of the reactor, respectively, as the time progresses.

The average pH in the anode chamber was 6.5 whereas an average pH of cathode chamber was 8.42 during stable period. The dissolved oxygen (DO) level in cathode chamber was found to be a major parameter for the biocathode. There is an average DO of 7.8 mg/L maintained at cathode which is sufficient to grow the aerobic bacteria. It was observed that anaerobic microbes at anode have self-buffering capacity, thus the pH at anode automatically comes to 7 without any external buffer addition. It was also observed that the high concentration of DO at cathode is very necessary for electron acceptor and also required for aerobic treatment of effluents. Oxygen was used as terminal electron acceptor in the cathode chamber because oxygen is freely available in the atmosphere and it does not produce any toxic by-products as the end product is water. The MFC has found to be a potential device for field application for effective wastewater treatment by the removal of organic matter from the wastewater. There were no other mediators like nitrate or permanganate solution added to the cathode chamber as an external agent for electron acceptor.

Conclusion

The current study on dairy wastewater treatment in the dual-chambered MFC employing low-cost earthenware separator demonstrated that the device is efficient in wastewater treatment with 89% COD removal efficiency and simultaneous electricity generation. The composite separator was efficient for cation exchange. The MFC employed with earthenware separator exhibited internal resistance of 108 Ω. The low-cost separator of 5 mm thickness has outperformed the most expensive Nafion membrane. The compositions of earthenware separator have confirmed the presence of silicon and aluminum which act as effective materials for proton exchange. The earthenware separator was found successful to withstand up to long period at such hydraulic load.

References

APHA, AWWA, WPCF (1998) Standard methods for examination of water and wastewater, 20th edn. American Public Health Association, Washington, DC

Behera M, Ghangrekar MM (2011) Electricity generation in low cost microbial fuel cell made up of earthenware of different thickness. Water Sci Technol 64(12):2468–2473

Chaudhuri SK, Lovely DR (2003) Electricity generation by direct oxidation of glucose in mediatorless microbial fuel cells. Nat Biotechnol 21(10):1229

Gorby YA, Beveridge TJ (2005) Composition, reactivity, and regulation of extracellular metal-reducing structures (nanowires) produced by dissimilatory metal reducing bacteria. Warrenton, VA

He Z, Minteer SD, Angenent LT (2005) Electricity generation from artificial wastewater using an upflow microbial fuel cell. Environ Sci Technol 39(14):5262–5267

Jana PS, Behera M, Ghangrekar MM (2010) Performance comparison of up-flow microbial fuel cells fabricated using proton exchange membrane and earthen cylinder. Int J Hydrogen Energy 35(11):5681–5686

Logan BE (2008) Microbial fuel cells. Wiley, Hoboken, NJ, USA

Rabaey K, Boon N, Siciliano SD, Verhaege M, Verstraete W (2004) Biofuel cells selectfor microbial consortia that self-mediate electron transfer. Appl Environ Microbiol 70(9): 5373–5382

Raj SA, Murthy DVS (1999) Synthetic dairy wastewater treatment using crossflow medium trickling filter. J Environ Sci Health Part a, 34(2):357–369

Zielińska M, Cydzik-Kwiatkowska A, Zieliński M, Dębowski M (2013) Impact of temperature, microwave radiation and organic loading rate on methanogenic community and biogas production during fermentation of dairy wastewater. Biores Technol 129:308–314

Part III
Environment: Chemical Environment

Synthesis of CaO_2 Nanoparticles for Environmental Remediation

Sapana S. Madan and Kailas L. Wasewar

Abstract Nanoparticles have important applications in the environment and remediation of contaminated soil and water. Calcium peroxide (CaO_2) nanoparticles signify a new generation of environmental remediation technology, which shows the cost-effective solutions for environmental cleanup problems. CaO_2 is used in nanosize by improving the high surface area, which can increase the speed of reaction. Nanoparticles of CaO_2 were synthesized by using chemical precipitation (surface modification) techniques. Synthesis of nanoparticles was confirmed by characterizing the sample using various techniques such as Fourier transform infrared (FTIR) spectroscopy, X-ray powder diffraction (XRD), high-resolution field emission scanning electron microscope (HR-FESEM), energy-dispersive X-ray spectroscopy (EDX), high-resolution TEM (HR-TEM). CaO_2 nanoparticles are a simple, low cost, ease to synthesize, economical and environment friendly. The synthesized nanoparticles can be used for the recovery of valuable chemicals from aqueous waste streams and also for removal of toxic materials.

Keywords Calcium peroxide · Nanoparticles · Remediation · Characterization

Introduction

The rapid speed of industrialization and their results have affected the environment by creating the harmful gases, fumes, and smokes in the environment. Nanotechnology proves to be of great interest in the environmental remediation. Nanoparticles have important applications in remediation of contaminated soil and water. Nanomaterials show enhanced performance in environmental remediation as

---

S.S. Madan · K.L. Wasewar (✉)
Advanced Separation and Analytical Laboratory, Department of Chemical Engineering, Visvesvaraya National Institute of Technology (VNIT), Nagpur 440010, Maharashtra, India
e-mail: k_wasewar@rediffmail.com

V.P. Singh et al. (eds.), *Energy and Environment*, Water Science and Technology Library 80, https://doi.org/10.1007/978-981-10-5798-4_7

compared to conventional techniques and it is used as an adsorbent, catalyst, and so on due to their high surface area (surface to volume ratio) and its high reactivity (Chaturvedi et al. 2012). Nanoparticles' sizes range in between 1 and 100 nm (Karn et al. 2009). The feasible result of nanoparticles on human health, animal, and the environment may be pleasant and unpleasant. Nanoparticles are used in various sectors such as biomedical, chemical, pharmaceutical, environmental, energy, electronic, cosmetic, magnetic, materials, opto-electronic, and catalytic (Biswas and Wu 2005). Nanomaterials combined with remediation show an effective approach than conventional methods because of the increased reactivity of nanoparticles and also possibility of in situ treatment (Mueller and Nowack 2010). Nanoremediation's possible effects are to reduce the overall costs of clean up and reduce the contaminant concentration close to zero and to prevent adverse environmental impacts. Nowadays, nanoparticles are used in water and wastewater treatment for removing pollutants such as heavy metals, acid recovery, and organic matter. Ground water is not pumped out for above-ground treatment in in situ method for nano remediation and soil, and it is also not transported to other places for treatment and removal (Otto et al. 2008). The number of nanoparticles has been developed for environmental remediation such as metal oxides, CNT, zeolites, and noble metals. The metal oxide has a number of research papers and very little work on inorganic peroxide. A great potential application of inorganic peroxides (chemical oxidants) related to organic synthesis, wastewater treatment, disinfection, and CO_2 adsorption (Berkessel and Vogl 2006). Metal peroxide compounds are used in industry as an oxidizing agent and as an oxygen source in organic synthesis because of their reactivity and oxidative capacity (Connor and Ebsworth 1964). Calcium peroxide (CaO_2) is a good choice for the remediation, but the speed of chemical oxidation between calcium peroxide and contaminant is slow. Calcium peroxide provides high stability in situ application and it has unique advantages in the environmental remediation. Calcium peroxide is slightly soluble in water. Calcium peroxide can be used in agriculture, aquaculture, and seed pretreatment (Ma et al. 2007). CaO_2 can cause skin irritation, affect eyes and respiratory tract. CaO_2 can be relatively used in the area of chemistry and biology such as wastewater purification, polymer chemistry, discharge of poisonous gases, yarn, bleaching of paper, water treatment, protection of plant from fungus infections, and terms of biology play the role of extra nutrition for birds and animals (Vol'nov 1983).

Conventional CaO_2 particles are excessively bulky (large) and because of this, its separation and movement will not be possible with the ground water flow to the environmental effect of remediation. It has been determined that it is used in nanosize and increases the surface area and reaction speed (Khodaveisi et al. 2011). It is used as a basis of chemical bound; however it can be simply evolved oxygen in fertilizers (Massalimov et al. 2010). It is used for oxygenation, disinfection of water, and in soil remediation (Qian et al. 2013). CaO_2 nanoparticles represent a new synthesis technology for environmental remediation to provide a cost-effective technique to the challenging environmental cleanup problems. In the present paper mainly focused on synthesis of CaO_2 nanoparticles which are inexpensive, low toxicity, easily separate from water, easy-to-scale-up so as to remove contaminants from water and wastewater. CaO_2 was confirmed by a variety of characterization techniques including Fourier transform infrared (FTIR) spectroscopy, X-ray powder diffraction (XRD), high-resolution field emission scanning electron microscope (HR-FESEM), energy-dispersive X-ray spectroscopy (EDX), transmittance electron microscopy (TEM), and high-resolution TEM (HR-TEM). Size determination by HR-TEM, TEM, and HR-FESEM indicate the size of synthesized CaO_2 nanoparticles is approximately 10–40 nm.

Experimental

Materials

All chemicals were obtained from Merck and used without any further purification. The analytical-grade chemicals $CaCl_2$ (99.5%), $NH_3{\cdot}H_2O$ (25%), PEG 200, H_2O_2 (35%), NaOH used for the synthesis of CaO_2 nanoparticles. Double-distilled water taken from double distillation unit (Remi India Ltd.) was used to prepare the aqueous solutions. A digital pH meter (Spectral Lab Instrumental Pvt. Ltd., India) was employed to measure the pH. 0.1 M NaOH was used to adjust the pH of the aqueous solutions.

Synthesis of CaO_2 Nanoparticles

In the current work, the synthesis procedure has been taken up (Khodaveisi et al 2011). The authors adapted synthesis of nanoparticles of zinc peroxide protocol (Rosenthal-Toib et al. 2008) and prepared nanoparticles of CaO_2 and the particle size range 15–25 nm, which was confirmed from their XRD and TEM results.

The ratio of synthesis procedure has been optimized with same conditions. 9 g of $CaCl_2$ were added in 1000 ml beaker with 90 ml double-distilled water. $NH_3{\cdot}H_2O$ solution (45 ml) and PEG 200 (360 ml) was added. H_2O_2 (45 ml) was added dropwise under stirring continuously for 2 h. The yellowish solution was formed after adequate time of stirring. NaOH solution was added until the pH becomes 11.5. The nanoparticles suspension was white in color and it was separated and collected by centrifugation at 10,000 rpm for 5 min. The centrifuge particles were washed three times with 0.1 N NaOH solutions and two times with double-distilled water. The pH 8.4 was obtained from the residue water. The centrifuged sample was dried in vacuum oven at 80 °C for 2 h. Figure 1 shows the schematic procedure for synthesis of CaO_2 nanoparticles.

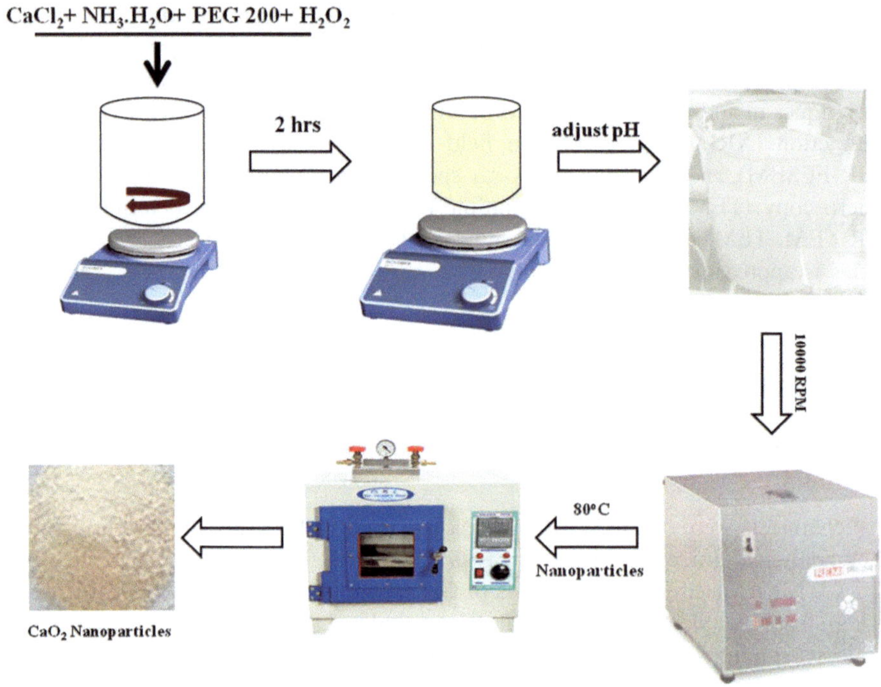

Fig. 1 Synthesis of nanoparticles of CaO_2

Characterization Techniques

Fourier transform infrared (FTIR) of CaO_2 were obtained using Perkin Elmer FTIR Spectrometer. X-ray diffraction (XRD) analysis was performed by X-ray diffractometer (PAN analytical X'pert PRO using Cu X-ray tube (λ = 1.5406 Å). Morphologies of samples were observed with a high-resolution field emission scanning electron microscope (HR-FESEM) from Zeiss, model name ULTRA Plus. Sample composition was quantified by energy-dispersive X-ray spectroscopy (EDX) (X Flash 6130, Bruker). High-resolution transmission electron microscopy (HR-TEM) of particles was carried out using JEOL JEM-2100.

Results and Discussion

During the present study, nanoparticles of calcium peroxide were obtained through chemical precipitation method using calcium chloride as precursor and hydrogen peroxide as oxidant. The synthesized nanoparticles of CaO_2 were characterized by various techniques. In the FTIR spectrum, CaO_2 was used to identify the functional

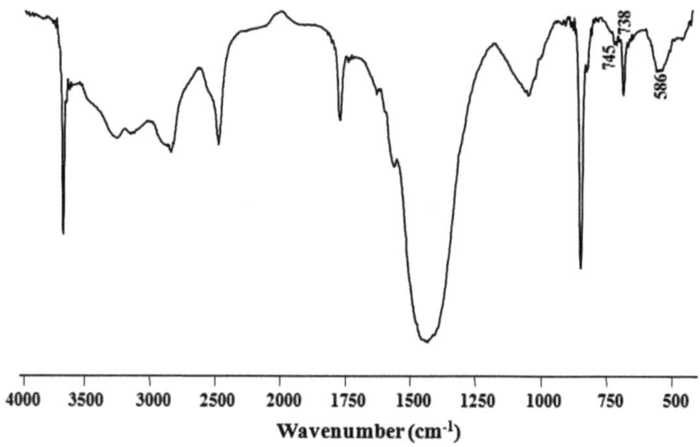

Fig. 2 FTIR of nano CaO_2

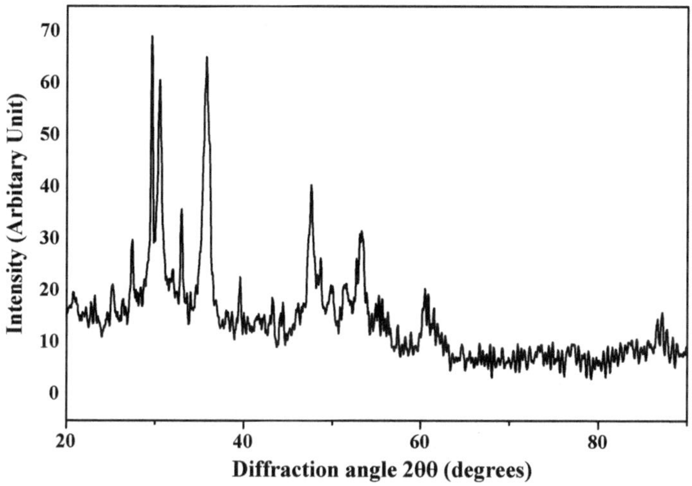

Fig. 3 XRD of nano CaO_2

groups present in nano CaO_2 (Fig. 2). The band at 586 cm^{-1} of CaO_2 corresponds to O–Ca–O vibration (Andrews et al. 1996). The band at 745 and 738 cm^{-1} of CaO_2 corresponds to the O–O bands of peroxide ions (Andrews et al. 1994).

The XRD pattern of the synthesized nano CaO_2 is shown in Fig. 3 which is in good agreement with that of the tetragonal structure of nano CaO_2. The sharp diffraction peaks indicate the preferential growth of crystallites of CaO_2 nanoparticles. The CaO_2 is identified from the XRD patterns by the peak positions at 30.47 (0 0 2), 35.75 (1 1 0), 47.46 (1 1 2), 53.28 (1 0 3), 60.82 (2 0 2), and 87.1 (3 1 0), which are in agreement with standard data.

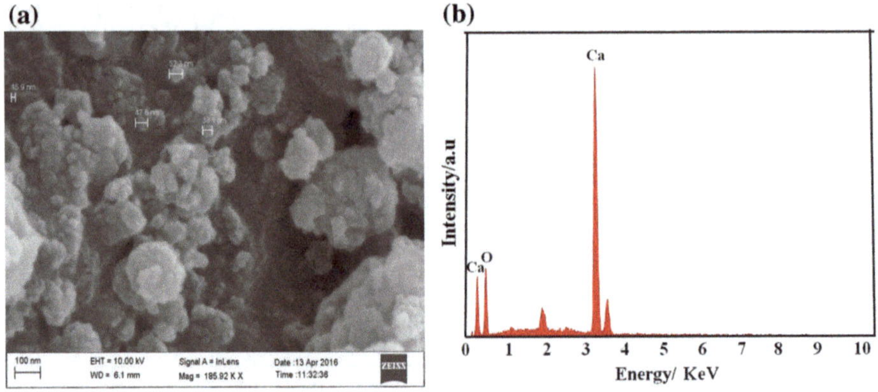

Fig. 4 **a** HR-FESEM and **b** EDX of nanoparticles of CaO_2

Figure 4a shows the HR-FESEM image of the prepared CaO_2 nanoparticles. HR-FESEM image of CaO_2 nanoparticles, which clearly indicates that nanoparticles were looked like the aggregated round shape and are mostly spherical in shape. As shown in Fig. 4a, average particle size is approximately 38.4 nm. Figure 4b shows the EDX spectra of synthesized CaO_2 nanoparticles. It determines the qualitative elemental composition of CaO_2 nanoparticles. Figure 4b shows the main peaks of calcium and oxygen and atomic % of Ca and O is about 1:2.

Additional, insight into the morphology of the synthesized nano CaO_2 was obtained by using HR-TEM image as shown in Fig. 5. HR-TEM analysis shows the single particle of nano CaO_2 with the particle size is 38 nm.

Environmental remediation incudes water remediation and the techniques used for this remediation are: adsorption, catalytic process, membrane process, radiation process, magnetically assisted process. From these different techniques, adsorption is largely used for removal and recovery from aqueous solution (Gopal et al. 2006; Reschke and Gelbin 1982). Nowadays, nanomaterials are a promising process for water, wastewater, and acid recovery treatment. Adsorption is a mass transfer phenomenon in which the substance is transferred from liquid phase to the surface of solid (Pollard et al. 1995). CaO_2 nanoparticles is used as an adsorbent in different streams for the removal of arsenic (Olyaie et al. 2012), toluene (Qian et al. 2013), and α-toluic acid (Madan et al. 2016). The removal efficiency of arsenic is 88% using very less CaO_2 nanoaprticle dose 40 mg/L. In our previous work, 97.7% removal efficiency of α-toluic acid is achieved using 5 g/L of CaO_2 nanoparticles dose (Madan et al. 2016). CaO_2 nanoparticles is efficiently used for environmental and water remediation.

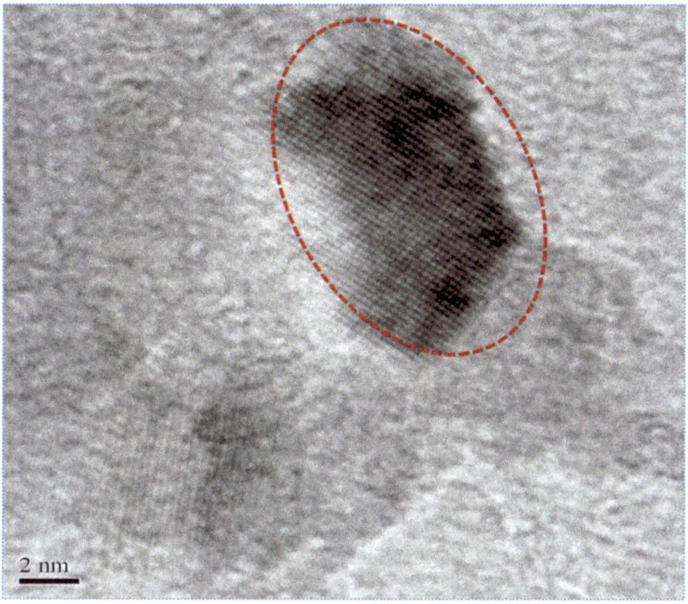

Fig. 5 HR-TEM of nano CaO_2

Conclusion

Environmental remediation has the potential to reduce the costs of cleaning up large-scale and to reduce contaminant concentrations in in situ treatment. The characterization of the CaO_2 nanoparticles was performed using Fourier transform infrared (FTIR) spectroscopy, X-ray powder diffraction (XRD), high-resolution field emission scanning electron microscope (HR-FESEM), energy-dispersive X-ray spectroscopy (EDX), high-resolution TEM (HR-TEM). From the results of HRSEM and HR-TEM, it was observed that the average particle size of CaO_2 nanoparticles is around 38 nm. CaO_2 nanoparticles can be successfully used for the environmental remediation of soil and ground water.

Acknowledgements We gratefully acknowledge SAIF NEHU Shillong for providing the HR-TEM facility; IISER Bhopal for carrying out the HR-FESEM facility; and Metallurgical & Materials Engineering Department, VNIT for providing the XRD facility.

References

Andrews L, Chertihin GV, Thompson CA, Dillon J, Byrne S, Bauschlicher CW Jr (1996) Infrared Spectra and quantum chemical calculations of group 2 MO_2, O_2MO_2, and related molecules. J Phys Chem 100:10088–10099

Andrews L, Yustein JT, Thompson CA, Hunt RD (1994) Reactions of pulsed-laser evaporated Ca, Sr, and Ba atoms with O_2. Infrared spectra of the metal oxides, oxide dimers, dioxides, and peroxides in solid argon. J Phys Chem 98:6514–6521

Berkessel A, Vogl N (2006) The chemistry of peroxides. In: Rappoport (ed). vol 2 Wiley, Ltd, Chichester (1), pp 307–586

Biswas P, Wu CY (2005) Nanoparticles and the environment. J Air Waste Manag Assoc 55: 709–746

Chaturvedi S, Dave PN, Shah NK (2012) Applications of nano-catalyst in new era. J Saudi Chem Soc 16(3):307–325

Connor JA, Ebsworth EAV (1964) Peroxy compounds of transition metals. Adv Inorg Chem Radiochem 6:279–381

Gopal R, Kaur S, Ma Z, Chan C, Ramakrishna S, Matsuura T (2006) Electrospun nanofibrous filtration membrane. J Membr Sci 281:581–586

Karn B, Kuiken BT, Otto M (2009) Nanotechnology and in situ remediation: a review of the benefits and potential risks. Environ Health Perspect 117(12):1823–1831

Khodaveisi J, Banejad H, Afkhami A, Olyaie E, Lashgari S, Dashti R (2011) Synthesis of calcium peroxide nanoparticles as an innovative reagent for in situ chemical oxidation. J Hazard Mater 192:1437–1440

Ma Y, Zhang BT, Zhao L, Guo G, Lin JM (2007) Study on the generation mechanism of reactive oxygen species on calcium peroxide by chemiluminescence and UV-visible spectra. Luminescence 22:575–580

Madan SS, Upwanshi AW, Wasewar KL (2016) Adsorption of α-toluic acid by calcium peroxide nanoparticles. Des Water Treat 57:16507–16513

Massalimov IA, Shayakhmetov AU, Mustafin AG (2010) Specific features of thermal decomposition of mechanically activated calcium peroxide. Russ J Appl Chem 83(10):1794–1798

Mueller NC, Nowack B (2010) Nanoparticles for remediation: solving big problems with little particles. Elements 6(6):395–400

Olyaie E, Banejad H, Afkhami A, Rahmani A, Khodaveisi J (2012) Development of a cost-effective techniqueto remove the arsenic contamination from aqueous solutions by calcium peroxide nanoparticles. Sep Purif Technol 95:10–15

Otto M, Floyd M, BajpaiS (2008) Nanotechnology for site remediation. Remediation 19(1): 99–108

Pollard SJT, Thompson FE, McConnachie GL (1995) Microporous carbons from Moringa oleifera husks for water purification in less developed countries. Water Res 29:337–347

Qian Y, Zhou X, Zhang Y, Zhang W, Chen J (2013) Performance and properties of nanoscale calcium peroxide for toluene removal. Chemosphere 91:717–723

Reschke G, Gelbin D (1982) Application of adsorption for water-purification: a literature-review. Chem Technol 34:114–120

Rosenthal-Toib L, Zohar K, Alagem M, Tsur Y (2008) Synthesis of stabilized nanoparticles of zinc peroxide. Chem Eng J 136:425–429

Vol'nov VI (1983) Perekisnyesoedineniyashchelochno-zemel'nykhmetallov (peroxide compounds of alkaline—Earth elements). Moscow, Nauka

Superiority of Re-circulating Fluidized Bed Reactor Over Existing Reactor Arrangements for Chemical Looping Combustion—A Review

Sachin Tomar, Nitin Lokachari and Raman Sharma

Abstract Chemical looping combustion (CLC) is an oxy-combustion CO_2 capture technique. In CLC, the fossil fuels combustion is carried out in two parts using metal oxides. First, the metal oxides are oxidized with air. Second, the metal oxides are reduced using the fuel. To fulfill this criterion of combustion for CLC, some researchers have used interconnected reactors arrangements and some have used single reactor. In interconnected reactors arrangement, metal oxides oxidation is carried out in one reactor and the reduction in other. However, in single reactor arrangement, the oxidation and reduction of metal oxide particles are carried out in cycles. The existing reactors arrangements suffer from the problems of high bed particle attrition, short residence time of the bed material/metal oxides in the air reactor (fast fluidized bed), and generation of fine particles, transfer of bed material from one reactor to another, complex loop-seal operation, heat losses during the transfer of bed material from one reactor to another and cluster formation. Re-circulating fluidized Bed (RCFB) reactor is proposed, here, to overcome some of the problems associated with the existing reactor arrangements. RCFB reactor has a concentric pipe arrangement. The inner pipe is known as riser and the outer pipe is known as downcomer.

Introduction

Scientists are working on a way to capture and sequester carbon dioxide to deal with global warming. Carbon Dioxide (CO_2) allows sunlight to reach the Earth and also prevents some of the sun's heat from radiating back into space, thus warming the planet (Metz et al. 2005). Yes, carbon dioxide and the greenhouse effect are necessary for Earth to survive. However, human inventions like power plants and transportation vehicles, which burn fossil fuels, release extra CO_2 into the air

S. Tomar · N. Lokachari · R. Sharma (✉)
Department of Chemical Engineering, Birla Institute of Technology and Science, Pilani 333031, Rajasthan, India
e-mail: raman.sharma@pilani.bits-pilani.ac.in

V.P. Singh et al. (eds.), *Energy and Environment*, Water Science and Technology Library 80, https://doi.org/10.1007/978-981-10-5798-4_8

thereby storing more amount of heat in the Earth's atmosphere. Carbon dioxide emissions alone increased by 80% in the past decade. Many researchers believe that the process of carbon capture and storage can help us to get this number down. Currently, there are many techniques readily available for carbon capture and storage (Haag 2007; Abad et al. 2007).

Chemical looping combustion (CLC) is a new technology that seems to work effectively for CO_2 capture. Carbon capture involves trapping the carbon dioxide at its emission source, transporting it to a remote location (usually deep underground) and isolating it (Mattisson et al. 2007). Initially, this process was proposed to increase the efficiency of thermal power plants; however, later on, it was found to have the inherent advantage of CO_2 capture avoiding costly equipment and energy consumption.

Chemical Looping Combustion (CLC)

CLC is an oxy-combustion CO_2 capture technique, which divides the process of combustion into two steps (Richter and Knocke 1983; Ishida et al. 1987). The first step takes place in an air reactor, where the oxidation of metal oxide particles using the compressed atmospheric air takes place (Eq. 1) (Lyngfelt et al. 2001; Leion et al. 2008; Mattisson et al. 2007) thereby stripping nitrogen from the air. The main exhaust gases from the air reactor are nitrogen and un-absorbed oxygen. After completion of this process, the oxidized metal particles or the oxygen carriers enter the fuel reactor, where the combustion takes place (Eq. 2) (Leion et al. 2008; Mattisson et al. 2007). Fuel gases are used to fluidize bed material in the fuel reactor which in the presence of oxygen completes the reduction cycle, thereby the combustion process (Mattisson et al. 2001). The main exhaust gases from the fuel reactor or the reduction cycle are CO_2 and H_2O. Once the water vapors are condensed, CO_2 can be separated easily (Leion et al. 2007; Mattisson et al. 2007).

$$\text{Air reactor (Oxidation cycle)}: \mathrm{Me} + 1/2\,\mathrm{O_2} \rightarrow \mathrm{MeO} \tag{1}$$

$$\text{Fuel reactor (Reduction cycle)}: (2n+m)\mathrm{MeO} + \mathrm{C}_n\mathrm{H}_{2m} \rightarrow (2n+m)\mathrm{Me} + m\,\mathrm{H_2O} + n\,\mathrm{CO_2} \tag{2}$$

Commonly used metal oxides are Fe, Ni, Mn, and Cu (Jin et al. 1998; Brereton and Grace 1993; Xu et al. 2009).

CLC Requirements

- Metal oxide particles (MeO) and fuel gas contact should be maximum.
- High residence time of the MeO particles in the fuel reactor.
- High solid circulation rate between the two reactors, i.e., air and fuel reactors.
- Uniform reactor temperature.

Existing Reactor Configuration for CLC

The most common interconnected fluidized bed configurations for CLC is shown in Fig. 1 in which air reactor is operated at high velocity and the fuel reactor is operated at low velocity. In this configuration, cyclone separator is used to retain the bed material in the system, which leave air reactor and diverted them to fuel reactor. The position of the fuel reactor is slightly elevated so that the bed material can easily flow from the fuel reactor to the air reactor under the influence of the gravity. Bed material circulation and flow of gases are governed by the loop-seals between the air reactor and the fuel reactor and between the cyclone separator and fuel reactor. Bed material circulates between the air reactor and the fuel reactor with the help of high velocity in the air reactor(Kolbitsch et al. 2009; Shen et al. 2009;

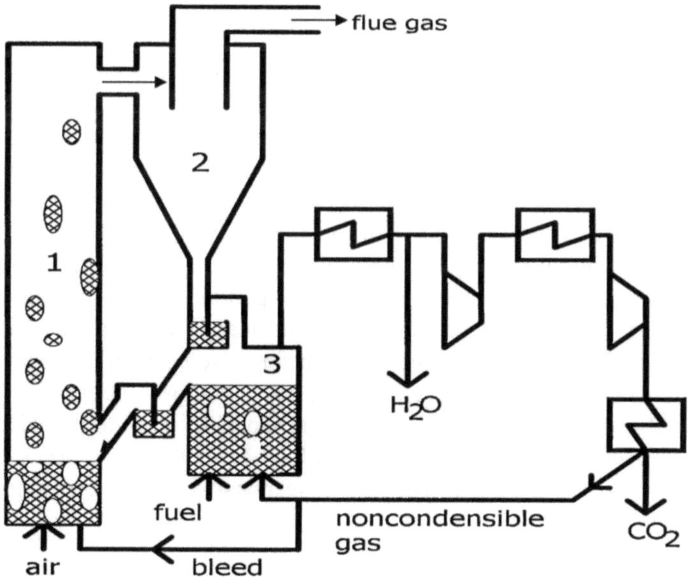

Fig. 1 Schematic diagram of interconnected reactor configuration for the CLC (Lyngfelt et al. 2001)

Basu and Cheng 2000; Johansson et al. 2003; Hossain and de Lasa 2008; Adanez et al. 2012).

Issues Related to the Existing Interconnected Reactors Arrangement

The main issues related to the existing interconnected fluidized bed reactors arrangements for chemical looping combustion (CLC), where air reactor is operated in a fast fluidization regime and the fuel reactor in bubbling fluidization regime are discussed here. The application of CLC principle for any of the fluidized bed systems mainly depends on the design of the fluidization column and the oxygen carrying particles (Merrow 1985). In order to overcome these problems, many modifications in the existing designs and many more new designs have been proposed. This paper mainly focuses on the following issues/drawbacks:

- Bed particle attrition
- Loop-seal operation
- Heat losses
- Cluster formation
- Fluidization gas bypassing
- Short residence time

Bed Particle Attrition

While moving in the reactor, the bed particles face attrition which is a mechanical stress due to inter-particle collision and bed-to-wall impacts. In the designing of the fluidized bed reactor, attrition is one of the main factors which are to be considered (Reppenhagen and Werther 2000). Due to the attrition, generation of fines take place which are finally passing through the dust recovery system resulting in the change in the properties of the bed material, particle size distribution gets too coarse, loss of valuable material, and wear and tear of the reactor (Bemrose and Bridgwater 1987). There are three main attrition sources in an interconnected reactor arrangement, namely, cyclone separator sections, the grid jet, and the bubbling bed as shown in Fig. 2. The cyclone separator attrition is higher than the jet and bubble-induced attrition at higher velocities. Equation 3 shows the overall loss rate as the sum of the individual sources responsible for the generation of fines.

$$\dot{m}_{\mathrm{loss,tot}} = \dot{m}_{\mathrm{loss},j} + \dot{m}_{\mathrm{loss},b} + \dot{m}_{\mathrm{loss},c} \tag{3}$$

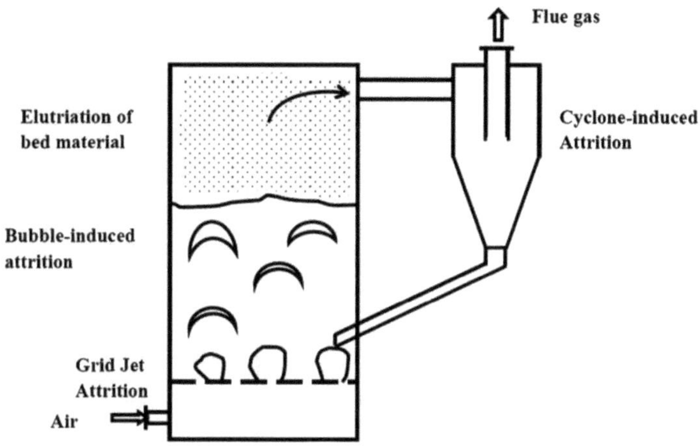

Fig. 2 Regions of bed particle attrition in an interconnected reactor configuration for CLC (Werther and Reppenhagen 1999)

where $\dot{m}_{\text{loss},j}$, $\dot{m}_{\text{loss},b}$ and $\dot{m}_{\text{loss},c}$ are the masses of elutriable fines that are produced per unit time by attrition in the jetting region, the bubbling bed, and the cyclone, respectively.

Loop-Seal Operation

Mechanical and nonmechanical valves are used for the transferring of the bed particles from one reactor to another reactor in the interconnected fluidized bed reactor system. Mechanical valves include rotary, screw, butterfly, and slide valves. These valves do not work properly under high temperature and pressure due to sealing and mechanical problem. At high temperature and pressure nonmechanical valves like loop-seal, *L*-, *J*-, *V*-valves (Knowlton and Hirsan 1978; Leung et al. 1987) are commonly employed to transfer bed particles from one reactor to another reactor by aeration as shown in Fig. 3. Loop-seal requires proper pressure balance around them, otherwise, seal failure may occur (Basu and Cheng 2000). Seal failure results in the gas bypassing which affects the overall efficiency of the CLC system. At a given gas velocity and solid circulation rate, the pressure drop across the loop-seal increases linearly with increasing solid inventory in the bed.

Heat Losses

In an interconnected fluidized bed reactor arrangement, there are many sites for the heat losses which include cyclone separator, loop-seal, transport line, standpipe, and reactor walls (Fig. 4) (Grace et al. 1987).

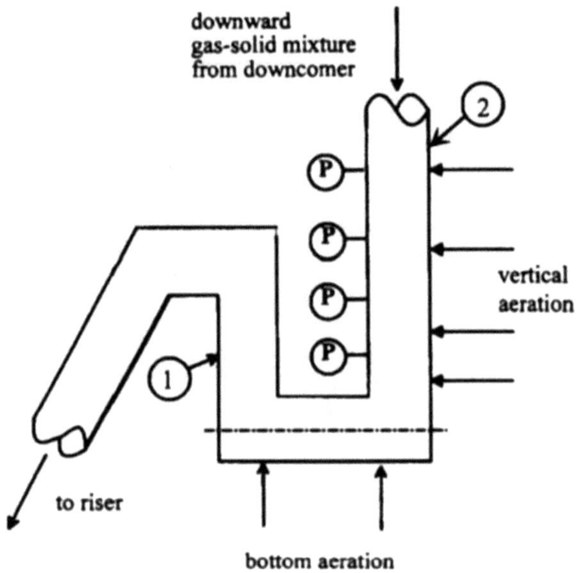

Fig. 3 Weir section (*1*) and vertical section of a loop-seal in an interconnected reactor configuration (*2*) (Kim et al. 1999)

Cluster Formation

The metal oxide particles used for carrying oxygen from air reactor to the fuel reactor have a great tendency to agglomerate or cluster formation, which is not a favored phenomenon in any of the fluidized bed operations. This might result in back-mixing of the particles, which affects the overall heat transfer rate and improper chemical interactions between metal oxide particles and the gas used for fluidization (Brereton and Grace 1993; Bai et al. 1995). When operated in a fast fluidization regime, riser or the draft tube section has very few upward moving particles and the downcomer section has many downward moving clusters (Chandel and Alappat 2006).

Gas Bypassing

The 'Gas' here refers to the one that is used for fluidization, i.e., air and fuel in the air reactor and the fuel reactor respectively. Gas bypassing can be defined as the escape of gas (air) from jet to downcomer, which is actually directed into the riser and vice versa (Kolbitsch et al. 2009; Chandel and Alappat 2005). This leads to escape of unburnt fuel from the fuel reactor resulting in improper combustion (Diana et al. 2011).

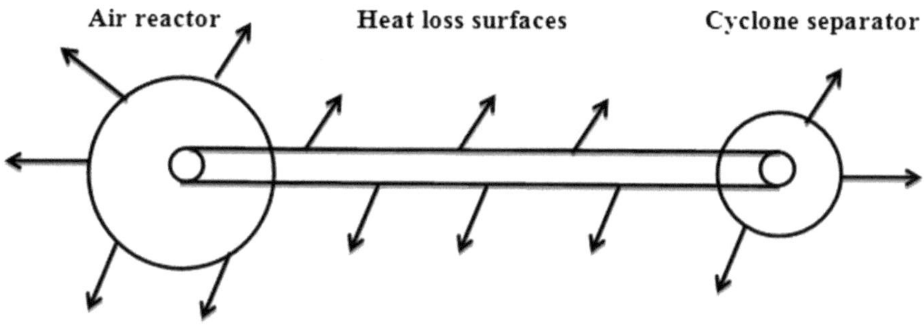

(a) Top View (heat loss from air reactor, transport line, and cyclone separator)

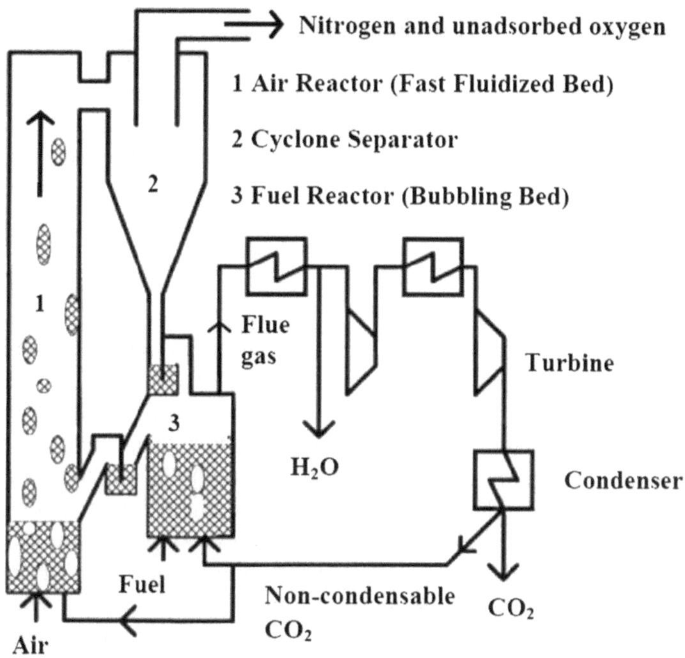

(b) Side view

Fig. 4 Heat loss surfaces in an interconnected fluidized bed reactor

Short Residence Time

Fast fluidization regime is preferred in the air reactor, which might affect the interactions of gas and the metal oxide particles, thus leading to a lower residence time. As a result of this, oxidation of metal oxide particles is improper which ends up in decreasing the performance of the reactor (Xu et al. 2009). In order to overcome the issues related to existing fluidized bed designs/setups, several modifications, and many new designs have been proposed by the researchers and one such design is the Re-circulating fluidized bed (RCFB) reactor. The reactor's design and its working have been briefly discussed here.

Re-circulating Fluidized Bed Reactor Design for CLC

RCFB has a concentric pipe arrangement, where the inner tube is called the draft tube and the outer or annular region is known as the downcomer. The fluidizing gas is supplied to the base of the draft tube through the air jet. RCFB can be compared to that of a spouted fluidized bed with a draft tube (Alappat and Rane 2001; Yang and Keairns 1978, 1983). RCFB eliminates the requirement for the cyclone separator, instead uses the freeboard section at the top. RCFB also consists of spacer section, whose length is changeable and the bed particles reside on the perforated plate. The principle of CLC can be applied to the RCFB design, where the inner draft tube acts as an air reactor and the downcomer as the fuel reactor.

Working

The fluidizing gas (compressed air) from the air jet is sent into the draft tube section, which comes into contact with the bed material, takes it along the draft tube (Alappat and Rane 2000; Yang and Keairns 1974) and in the mean-time the particles get oxidized and enter the freeboard section, lose their energy and fall back into the downcomer section, where the fluidizing gas will be any fossil fuel, reduces the metal oxide particles. Though the process of combustion is divided into two parts, the main part remains to take place in the downcomer section i.e. fossil fuel reacting with the oxidized metal particles, producing energy. These particles again enter the bottom of the draft tube to start a new oxidation cycle. The schematic representation of RCFB is explained with the real-time cold model setup available in BITS Pilani (Fig. 5).

How RCFB overcomes the issues related to the existing designs? A design by the name RCFB has been introduced which can possibly overcome some of the issues related to the existing reactor arrangements. The construction of this particular fluidized bed is done in such a way, keeping in mind the sole purpose, i.e., to overcome the existing problems.

- As cyclone separator is the main site of attrition at the higher velocity, so in RCFB reactor a large cross-sectional area called freeboard section is used in place of cyclone separator to retain the bed particles in the reactor which results in lesser particle attrition than the existing reactor configuration, without losing the bed particles.

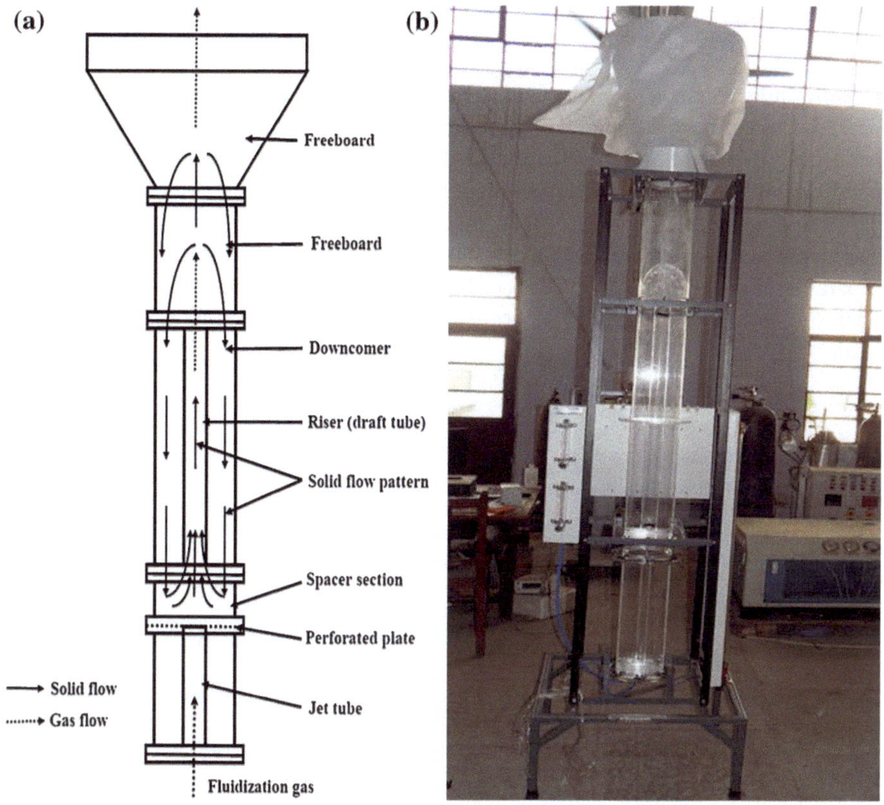

Fig. 5 **a** Schematic diagram of RCFB reactor. **b** Lab scale RCFB experimental setup in BITS Pilani

- The complex loop-seal arrangement is not needed in the RCFB reactor which prevents gas bypassing.
- In RCFB reactor configuration, cyclone separator and complex loop-seal arrangement are not used and hence less heat loss.
- The draft tube also induces high solid circulation rate and bed particles keep circulating within the reactors, which ensures no cluster formation.
- The extra travel path provided by the draft tube will ensure more contact time between the particles and the gas (Yang and Keairns 1978).
- Maintaining a large bed inventory in the spacer section and also the lesser length of the spacer section will not allow the gas to bypass.

Conclusions

Existing interconnected reactor arrangements for CLC have some issues that have been discussed in this paper. To overcome some of these issues, RCFB reactor has been proposed here. The draft tube in the RCFB ensures good solid mixing and solid circulation, longer residence time which results in maintaining uniform temperature throughout the reactor and better distribution of bed inventory and fuels. The construction of the RCFB reactor is not complex as it does not have cyclone separator and complex loop-seals, which makes it less expensive with flexible operating conditions. Further experimental studies are needed to verify the claims made.

References

Abad A, Mattisson T, Lyngfelt A, Johansson A (2007) The use of iron oxide as an oxygen carrier in a chemical-looping reactor. Fuel 86:1021–1035

Adanez J, Abad A, Labiano AG, Gayan P, de Diego LF (2012) Progress in chemical-looping combustion and reforming technologies. Prog Energy Combust Sci 38:215–282

Alappat BJ, Rane VC (2000) Performance prediction of an RCFB incinerator system. Energy Eng 126:53–65

Alappat BJ, Rane VC (2001) Solid circulation rate in recirculating fluidized bed. J Energy Eng 127:51–68

Bai D, Shibuya E, Masuda Y, Nishio K, Nakagawa N, Kato K (1995) Distinction between upward and downcomer flows in circulating fluidized beds. Powder Technol 84:75–81

Basu P, Cheng L (2000) An analysis of loop seal operation in a circulating fluidized bed. Trans Inst Chem Eng 78(A):991–998

Bemrose CR, Bridgwater J (1987) A review of attrition and attrition test methods. Powder Technol 49:97–126

Brereton CMH, Grace JR (1993) Microstructural aspects of the behavior of circulating fluidized bed. Chem Eng Sci 49(14):2565–2572

Chandel MK, Alappat BJ (2005) Pressure drop and gas by-passing in recirculating fluidized beds. Chem Eng Sci 61:1489–1499

Chandel MK, Alappat BJ (2006) Annular down flow layer in a recirculating fluidized bed. Ind Eng Chem Res 45:5748–5754

Diana C, Pérez G, Tondl G, Höltl W, Pröll T, Hofbauer H (2011) Cold flow model study of an oxyfuel combustion pilot plant. Chem Eng Technol 34(12):2091–2098

Grace JR, Lim CJ, Brereton CMH, Chaouki J (1987) Circulating fluidized bed reactor design and operation. Sadhana 10(1&2):35–48

Haag AL (2007) Post-Kyoto pact: shaping the successor. Nat Rep Clim Change 1:12–15

Hossain M, de Lasa HI (2008) Chemical-looping combustion (CLC) for inherent CO_2 separations—a review. Chem Eng Sci 63:4433–4451

Ishida M, Zheng D, Akehata T (1987) Evaluation of a chemical looping-combustion power-generation system by graphic exergy analysis. Energy 12(2):147–154

Jin H, Tsutsumi A, Yoshida K (1998) Solids circulation in a spouted fluid bed with the draft tube. Can J Chem Eng 31(5):842–845

Johansson E, Lyngfelt A, Mattisson T, Johnsson F (2003) Gas leakage measurements in a cold model of an interconnected fluidized bed for chemical-looping combustion. Powder Technol 134:210–217

Kim SW, Namkung W, Kim SD (1999) Solids flow characteristics in loop-seal of a circulating fluidized bed. Korean J Chem Eng 16(1):82–88

Knowlton TM, Hirsan I (1978) L-valves characterized for solid flow. Hydrocarb Process 57:149

Kolbitsch P, Pröll T, Kampf BNJ, Hofbauer H (2009) Design of a chemical looping combustor using a dual circulating fluidized bed reactor system. Chem Eng Technol 32(3):398–403

Leion H, Mattisson T, Lyngfelt A (2007) The use of petroleum coke as fuel in chemical-looping combustion. Fuel 86:1947–1958

Leion H, Mattisson T, Lyngfelt A (2008) Solid fuels in chemical-looping combustion. Int J Greenh Gas Control 2:180–193

Leung LS, Chong YO, Lottes J (1987) Operation of V-valves for gas-solid flow. Powder Technol 49:271

Lyngfelt A, Leckner B, Mattisson T (2001) A fluidized bed combustion process with inherent CO_2 separation; application of chemical-looping combustion. Chem Eng Sci 56:3101–3113

Mattisson T, Lyngfelt A, Cho P (2001) Theuse of iron oxide as an oxygen carrier in chemical-looping combustion of methane with inherent separation of CO_2. Fuel 80:1953–1962

Mattisson T, Labiano GF, Kronberger B, Lyngfelt A, Adánez J, Hofbauer H (2007) Chemical-looping combustion using syngas as fuel. Int J Greenh Gas Control 1:158–169

Merrow EW (1985) Linking R&D to problems experienced in solids processing. Chem Eng Process 81(5):14–22

Metz B, Davidson O, de Coninck H, Loos M, Meyer L (2005) Carbon dioxide capture and storage. Cambridge University Press, New York

Reppenhagen J, Werther J (2000) Catalyst attrition in cyclones. Powder Technol 111:55–69

Richter H, Knocke K (1983) Reversibility of combustion processes. Am Chem Soc Symp Ser 235:71–86

Shen L, Wu J, Xiao J (2009) Experiments on chemical looping combustion of coal with a NiO based oxygen carrier. Combust Flame 156:721–728

Werther J, Reppenhagen J (1999) Catalyst attrition in fluidized-bed systems. React Kinet Catal 45 (9):2001–2010

Xu M, Ellis N, Lim CJ, Ryu HJ (2009) Mapping of the operation conditions for an interconnected fluidized bed reactor for CO_2 separation by chemical looping combustion. Chem Eng Technol 32(3):404–409

Yang WC, Keairns DL (1974) Recirculating fluidized-bed reactor data utilizing a two-dimensional cold model. Am Inst Chem Eng Symp Ser 70(141):27–40

Yang WC, Keairns DL (1978) Design of recirculating fluidized beds for commercial applications. Am Inst Chem Eng Symp Ser 74(176):218–228

Yang WC, Keairns DL (1983) Studies on the solid circulation rate and gas bypassing in spouted fluid-bed with a draft tube. Can J Chem Eng 61(June):349–355

Enhancing N Use Efficiency and Reducing N_2O Emission by Coating Urea with Newly Identified Bio-Molecule ($C_{20}H_{30}O_2$), Nano-Zn Oxide and Nano-rock Phosphate

S. Kundu, Tapan Adhikari, M. Vassanda Coumar, S. Rajendiran and J.K. Saha

Abstract The increasing use of nitrogenous fertilizer and its importance in Indian agriculture has attracted the attention of environmentalists. There is a greater concern that nearly 30–60% of applied N is lost in the form of ammonia from the agricultural soils through volatilization. To reduce such loss, urea can be coated with pine oleoresin, a novel coating material, for making slow release urea with minimal cost using the solvents like petrol. The POR-coated urea releases N slowly by the action of a physical barrier and increase N availability by inhibiting urease activity through antibacterial properties and reduces volatilization loss by acidifying alkaline micro-sites in soil. It also improves the biomass yield, nitrogen uptake and nitrogen use efficiency. Keeping in view of the above, the current investigation was aimed to coat urea with nano-ZnO (<100 nm) and nano-rock phosphate (<48.6 nm) using pine oleoresin (POR) as a coating agent and to study their effect on nitrous oxide (N_2O) emission. Significant improvement in N use efficiency ranging 10–20% was observed in greenhouse studies. Further, the results clearly demonstrated that further encapsulation of POR coated urea with nano-Zn oxide (<50 nm) and nano-rock phosphate (<100 nm) reduced nitrous oxide emission by 30–40%. Nano-ZnO coated urea showed the least N_2O emission (0.28 μg N_2O mg^{-1} N) followed by 35% nano-RP coated urea (0.30 μg N_2O mg^{-1} N). The results thus indicated that coating urea with 2% nano-ZnO and 35% nano-rock phosphate could be used as a potential means to reduce N_2O emission from fertilizer N.

S. Kundu (✉) · T. Adhikari · M. Vassanda Coumar · S. Rajendiran · J.K. Saha
Indian Institute of Soil Science, Nabibagh, Berasia Road, Bhopal 462038, India
e-mail: samareshk_2006@yahoo.com

V.P. Singh et al. (eds.), *Energy and Environment*, Water Science and Technology Library 80, https://doi.org/10.1007/978-981-10-5798-4_9

Introduction

Urea is the most important commercial fertilizer in India and currently 13.7 million MT of N is used in Indian agriculture through the application of urea. Efficient use of N is absolutely essential for sustaining agricultural productivity as well as acting as a driving force for climate mitigation. Because of strong inter-relationship among N uses, climate change and food security, there is an urgent needs to search possibilities to increase N use efficiency in crop production and to reduce N_2 emission from applied fertilizer N. Worldwide, use efficiency of N from fertilizer is as low as 30–35% (Galloway and Cowling 2002; Ladha et al. 2005). In Indian context, the N use efficiency of fertilizer N (%N recorded by crops) is only 30–50% and it is as low as 20% in rice (Prasad 1998). Globally, it has been estimated that 10–30% of the externally applied N is used in the end products (Smil 2001) and rest is returned to the ecosystem in the form of gases (such as N_2, N_2O, NH_3 and NO) or leaked to the water bodies as soluble NO_3.

Today, Global warming is the most prominent environmental issue before the humanity. It is caused by the increase in concentration of greenhouse gases (GHGs) in the atmosphere. The major sources are soil cultivation, fertilizer and manure application, and burning of crop residues. Application of N fertilizer is the major source of N_2O emission contributing 70%, followed by crop residue burning and mineralization of organic N in soil (Pathak and Aggrawal 2012). Unlike emission of CH_4 which has remained constant over the years, emission of N_2O has been increasing with the increase in fertilizer N use in Indian agriculture. The most efficient management practice to reduce N_2O emission is site-specific, efficient nutrient management (Pathak et al. 2004). The emission could also be reduced by nitrification inhibitors such as nitrapyrin and dicyandiamide (DCD). There are some plant-derived organic such as neem oil, neem cake and karanja seed extact which also act as nitrification inhibitors.

In our earlier experiment (Kundu et al. 2013), we demonstrated the efficacy of newly identified bio-molecule pine oleoresin ($C_{20}H_{30}O_2$) as a potential urease inhibitor and coating material for making slow release urea. Pine oleoresin (POR) is commercially extracted from pine tree (*Pinus roxburghii*) and is having empirical formula $C_{20}H_{30}O_2$. The natural POR contains four resin acids, namely, levopimaric acid (22%), palustric acid (11%), I-abietic acid (10%) and neoabietic acid (15%). All these four acids comprise about 50–60% of the acidic portion of the POR. This group of conjugated-diene and acids are frequently referred to as abietic-type acids (Baldwin et al. 1958). Of the four conjugated-diene resin acids, levopimaric acid possesses the most reactive system of double-bonds and is more resistant to oxidation (Lloyd and Hedrick 1965). Coating urea with POR provides a physical barrier for slow release of N from urea, inhibits urease activity through antibacterial properties and reduces volatilization loss by acidifying alkaline micro-sites in soil (Kundu et al. 2013).

In recent years remarkable progress has also been made in developing nanotechnology and it also provides the tool and the technological platforms to investigate its effect on biological systems. To address the issues relating to increase fertilizer use efficiency nanotechnology in agricultural sector has been introduced, which is likely to bring a sea change in the production of fertilizers, which are expected to improve agricultural production and productivity. Keeping in view of all the above, the current investigation was aimed to coat the urea with pine oleoresin, nano-ZnO and nano-rock phosphate (RP) materials and to study their effect on crop yield, nutrient use efficiency and nitrous oxide (N_2O) emission from applied urea in soil.

Materials and Methods

Protocol for the Preparation of Nano-ZnO and Nano-rock Phosphate Coated Urea

In our previous experiment (Kundu et al. 2013), an attempt was made to coat urea with pine oleoresin that provides a physical barrier for slow release of N from coated urea; inhibits urease activity through antibacterial properties; and reduces volatilization loss by acidifying alkaline micro-sites. The protocol for making POR coated slow release urea involved the use of hexane as a solvent. In search of suitable cheaper solvent to dissolve POR, we found that commercial petrol could be used as a solvent in place of hexane. Protocol for the preparation of nano-ZnO and nano-rock phosphate coated urea involved dissolution of 20% POR in commercial petrol. The requisite amount of Urea (1 kg) was mixed with above solvent (250 ml) in the ratio of 4:1 in wide mouth glass bottle and shaken for 5 min. Immediately after mixing, the whole content was transferred to a plastic tray fitted snugly on a horizontal shaker and shaking operation was continued with maximum speed for an hour, with intermittent scrubbing using a hard brush. During the second operation, 20 g nano-ZnO, 150, 200, 250, 300 and 350 g nano-rock phosphate were sprayed over the POR mixed urea. After the complete evaporation of solvent, the nano-ZnO and nano-rock phosphate coated urea became loose and friable and thereafter kept in oven (50 °C) for overnight for hardening. The nano-ZnO (<100 nm) was obtained from Sigma-Aldrich (Photo 1) and the nano-rock phosphate (<48.6 nm) was prepared from Udaipur rock phosphate (34% P_2O_5) using high energy ball-mill (Photo 2). Following this protocol we obtained seven different types of coated urea (Table 1 and Photo 3). The nutrient content in these fertilizer materials is presented in Table 1.

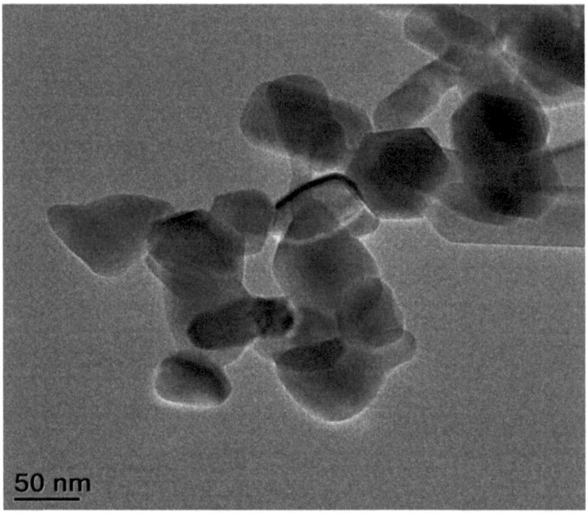

Photo 1 SEM image of nano-ZnO particles

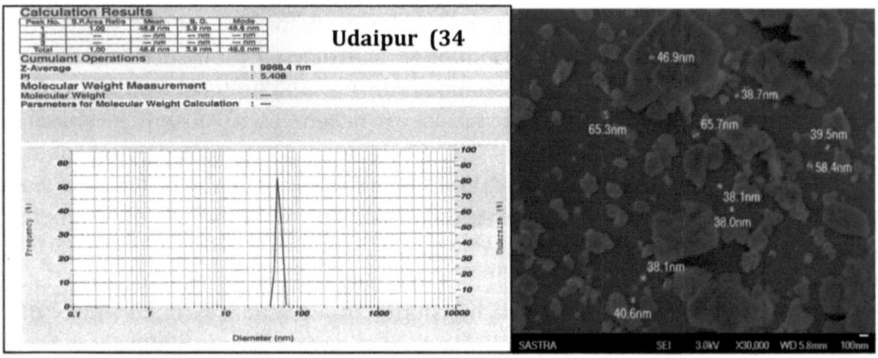

Photo 2 Particle size distribution and SEM image of nano-rock phosphate particles

Table 1 The nutrient content of the different coated urea fertilizers

Fertilizer material	Nutrient content (%)		
	N	P_2O_5	Zn
Urea (F_1)	46.00	–	–
Urea coated with 5% POR (F_2)	43.80	–	–
Urea coated with 5% POR and 2% nano-ZnO (F_3)	42.99	–	1.45
Urea coated with 5% POR and 15% nano-RP (F_4)	38.33	4.25	–
Urea coated with 5% POR and 20% nano-RP (F_5)	36.80	5.44	–
Urea coated with 5% POR and 25% nano-RP (F_6)	35.38	6.53	–
Urea coated with 5% POR and 30% nano-RP (F_7)	34.07	7.55	–
Urea coated with 5% POR and 35% nano-RP (F_8)	32.85	8.50	–

Photo 3 Coated urea fertilizers. **a** POR-coated urea; **b** 15% nano-RP fortified POR-coated urea; **c** 20% nano-RP fortified POR-coated urea; **d** 25% nano-RP fortified POR-coated urea; **e** 30% nano-RP-fortified POR-coated urea; **f** 35% nano-RP fortified POR-coated urea

Table 2 Characteristics of pot culture experiment soils

Characteristics	Vertisol	Alfisol
pH (1:2.5 H_2O)	7.85	5.65
EC (dS m^{-1})	0.4	0.35
SOC (%)	0.66	0.40
Clay (%)	55.5	24.9
CEC (cmol (p^+) kg^{-1})	45.0	8.0
Available N (kg ha^{-1})	341	263
Available P (kg ha^{-1})	13.2	10.3
Available K (kg ha^{-1})	386	292

Pot Culture Experiment on Crop Yield and Nutrient Use Efficiency

A pot culture experiment (10 kg capacity pot) was conducted during Kharif season of 2014 to study the uptake of N and P by maize crop grow on two soils (Alfisol from Betul, M.P. and Vertisol from IISS, Bhopal farm) as affected by the application of urea fortified with variable amounts of nano-rock phosphate (varying from 15–35% on mass basis) using pine oleoresin as a coating agent. The characteristics of the soils are given in Table 2. There were eight treatments, namely, absolute control without N (T_1), uncoated urea + P(SSP) (T_2), 5% POR coated urea + P (SSP) (T_3), 5% POR coated urea + 15% RP (T_4), 5% POR coated urea + 20% RP (T_5), 5% POR coated urea + 25% RP (T_6), 5% POR coated urea + 30% RP (T_7)

and 5% POR coated urea + 35% RP (T_8). The treatments were imposed in completely randomized design with four replications. The required amounts of uncoated urea and coated urea were mixed with the top 50% soil of the pots. While, mixing urea, 100 ml of KH_2PO_4 solution was mixed with the soil so as to supply 40 ppm P and 50 ppm K for each pot. After maintaining moisture content at field capacity, maize seeds (var. Kanchan) were sown @ 5 seed per pot and after germination, finally 3 plants were maintained. The crop was grown up to 55 days and thereafter harvested, oven dried and biomass yields were recorded. The N content in the biomass was estimated by standard procedure. The N use efficiency from different coated and uncoated urea was measured by the following equation:

$$\text{NUE}\ (\%) = \frac{\text{N uptake in treated pot} - \text{N uptake in control pot}}{\text{Amount of N applied}} \times 100.$$

Experimental Details and Measurement of N_2O Emission

The N_2O emission from these fertilizer materials (Table 1) was measured in the laboratory (at 22 °C) using 125 ml glass bottle with an air tight stopper fitted with silicon septum. In each bottle 10 g rhizosphere soil (Soil type: Vertisol; pH: 7.9; clay: 55.5%) was placed and different fertilizer material was mixed so as to supply 9.2 mg N uniformly. After mixing fertilizer N, 4 ml distilled water was added and the bottle was sealed. The head space for collection of N_2O gas in each bottle was 120 ml. For each fertilizer material, 13 such bottles were maintained, of which 10 bottles were used for periodic measurement and remaining 3 bottles were used for total emission after 33 days. Also, 13 bottles with soil (without fertilizer) were maintained to measure the emission of N_2O from the test soil. The gas sample was drawn with 1 ml hypodermic needle at different time intervals up to 33 days. During periodic measurement of N_2O emission, 3 gas samples were collected from each bottle on each day. The concentration of N_2O in the head space was analyzed by Gas Chromotograph (model: CIC make GC 2010) fitted with an electron capture detector and 6′ × 1/8″ stainless steel column (Poropak N). GC software was used to plot and measure the peak area using suitable N_2O gas standard. The average peak area was 27 for 10 μl/L or 10 ppm V N_2O. The N_2O concentration measured by GC was converted to mass basis using ideal gas law equation:

$$PV = nRT$$

where, P = pressure (0.965 atmosphere), V = Volume of gas (L), n = number of moles of gas, R = gas law constant (0.08206 L atm Mol^{-1} K^{-1}), and T = temperature (295 °K). From the above expression we calculated that 1 μL N_2O gas (at 1000 feet altitude and 22 °C) contained 0.0398 μmol N_2O or 1.7539 μg N_2O.

Statistical Analysis

Samples were analyzed in triplicate and mean values were used for comparisons. Variance analysis was performed with SPSS (Version 10.0) statistical software. Means were compared by the critical difference (CD) test ($P \leq 0.05$).

Results and Discussion

Nutrient Content of Coated Fertilizers

When urea was coated with 5% POR, N content of the urea was decreased from 46.0 to 43.8% in POR coated urea and further it decreased to 42.99% due to addition of 2% nano-ZnO to the coating. However POR plus nano-ZnO coated urea had an advantage of 1.45% Zn in its content. Further POR coated urea was inculcated to hold 15, 20, 25, 30, and 35% of nano-RP and N content decreased from 43.8 to 38.33, 36.80, 35.38, 34.07 and 32.85%, respectively. Though N content decreased in the POR plus nano-RP urea fertilizers, P content of the fertilizers (in terms of P_2O_5) increased to 4.25, 5.44, 6.53, 7.55 and 8.50%, respectively (Table 1). Hence it had benefits of both the nutrients N and P, and may be used as slow release complex fertilizer.

Impact of Urea Coated with Pine Oleoresin, Nano-Zn Oxide and Nano-rock Phosphate on Crop Yield and Nutrient Use Efficiency

The biomass yield of maize after 55 days of sowing on two different soils treated with N supplied through uncoated urea and urea coated with 2% nano-ZnO and different amount of nano-RP is presented in Tables 3 and 4. In both the soils, application of N @138 ppm as uncoated urea resulted significant increase in biomass yield of maize to the tune of 72.6 and 68.2%, in Alfisol and Vertisol, respectively. Further, application of equivalent amount of N through urea coated with increasing amount of nano-RP coating (ranging from 15 to 35% RP) showed increasing trend in biomass yield of maize in both the soils. Further, application of urea coated with 5% POR in Alfisol significantly increased the biomass yield by 8.7% over normal urea (uncoated urea) in Alfisol. However in Vertisol, significant increase in biomass yield was not observed in the urea coated with 5% POR over normal urea. It has been reported that coated urea can produce a high yield due to the slow release of nitrogen (Junejo et al. 2011; Purkayastha and Chhonkar 2006). The biomass increase in the coated urea treated pots may be due to sustained release and increased availability of N for the crop by slow release of N from coated urea,

Table 3 Biomass yield of maize as affected by application of nano-rock phosphate coated urea in red soil

Treatment	Yield (g/pot)	%N in grain	%P in grain	Grain N uptake (mg/pot)	Grain P uptake (mg/pot)	NUE (%)	PUE (%)
Control	27.87	0.651	0.214	181.43	59.64	–	–
U + P	48.11	1.226	0.248	589.82	119.31	29.59	38.87
CU + P	52.29	1.243	0.251	649.96	131.24	33.95	46.64
CU + 15% RP	39.58	1.217	0.196	481.68	77.57	21.75	27.26
CU + 20% RP	42.38	1.106	0.220	468.72	93.23	20.82	38.30
CU + 25% RP	42.88	1.152	0.224	493.97	96.05	22.64	33.25
CU + 30% RP	44.15	1.205	0.229	532.00	101.10	25.40	31.53
CU + 35% RP	46.38	1.192	0.241	552.84	111.77	26.91	33.96
LSD ($p = 0.05$)	3.27	0.048	0.023	47.40	11.12	2.53	3.28

U Normal urea, *CU* POR coated urea, *P* SSP@153.51 mg P/pot, *N dose* 1380 mg/pot
NUE Nitrogen use efficiency, *PUE* Phosphorus use efficiency

Table 4 Biomass yield of maize as affected by application of nano-rock phosphate coated urea in black soil

Treatment	Yield (g/pot)	%N in grain	%P in grain	Grain N uptake (mg/pot)	Grain P uptake (mg/pot)	NUE (%)	PUE (%)
Control	32.28	0.682	0.196	220.15	63.26	–	–
U + P	54.30	1.248	0.236	677.66	128.15	33.15	42.27
CU + P	56.15	1.251	0.241	702.45	135.32	34.95	46.94
CU + 15% RP	43.53	1.213	0.203	528.01	88.36	22.31	38.16
CU + 20% RP	45.12	1.218	0.214	549.56	96.55	23.87	37.95
CU + 25% RP	46.75	1.220	0.208	570.35	97.24	25.37	31.03
CU + 30% RP	50.18	1.206	0.217	605.17	108.89	27.90	34.70
CU + 35% RP	51.97	1.216	0.223	631.95	115.89	29.84	34.28
LSD ($p = 0.05$)	3.92	0.034	0.029	24.84	9.69	2.68	3.62

U Normal urea, *CU* POR-coated urea, *P* SSP@153.51 mg P/pot, *N dose* 1380 mg/pot, *NUE* Nitrogen use efficiency, *PUE* Phosphorus use efficiency

inhibits urease activity through antibacterial properties and reduces volatilization loss by acidifying alkaline micro-sites in soil. Further, the result also showed that the urea coated with varying amount of RP (15–35%) significantly reduced the biomass yield as compared to coated urea without RP. The reduction in biomass yield due to RP coating might be due to low N content in the coated urea coupled with its slow release of nitrogen from the RP coated urea.

The N content in maize grain was improved significantly both soils due to application of 138 ppm N through uncoated urea as well as urea coated with POR and different amount of RP (Tables 3 and 4). Coating urea with natural materials is an effective method of reducing urea hydrolysis. Slow hydrolysis allows urea to remain in fertilized pots for long period of time due to the reduced loss of ammonia via volatilization and caused by high amounts of ammonium accumulation on fertilizer microsite. In absence of external N application, N uptake in grain by the 55 days old maize crop varied widely, ranging from 181.4 mg/pot in Alfisol to 220.1 mg/pot in Vertisol (Tables 3 and 4), depending upon the native fertility status of the soils. Because of the application of external N @138 ppm through uncoated urea, the total N uptake increase by 225.1, and 207.8%, in Alfisol and Vertisol, respectively, over control. Further, urea coated with 5% POR significant increased the grain N uptake from 589.82 to 649.96 mg/pot in Alfisol, while in Vertisol, the grain N uptake was from 677.66 to 702.45 mg/pot as compared to uncoated urea. However, coating urea with RP of varying quantity (15–35%) significantly reduced grain N uptake both in Alfisol and Vertisol as compared to urea coated with 5% POR only (without RP). Coating enhanced the N availability of urea fertilizer and reduced ammonia loss by reducing soil pH and ammonium accumulation on the soil surface that resulted in increased uptake of N by the crop in all the soils under study. However, N uptake remained low at uncoated urea applied soils, which indicates high ammonia loss.

The N use efficiency (i.e. ratio of N harvested from fertilizer N to N applied as fertilizer) of urea fertilizer increased remarkably, ranging from 20.8 to 33.9% in Alfisol and 22.3 to 34.9% in Vertisol due to coating of urea with POR and varying amount of RP (Tables 3 and 4). Application of urea coated with 5% POR showed significant improvement in N use efficiency over uncoated urea in both Vertisol and Alfisol, but urea coated with POR and varying level of RP significantly reduced the NUE both in Alfisol and Vertisol. Maximum N use efficiency in Alfisol (33.9%) and Aridisol (34.9%) was observed with urea coated with 5% POR. The results showed that under uniform dose of N (138 ppm), N uptake and N use efficiency improved due to coating with POR. Coating urea with increasing amount of nano-rock phosphate decreased N uptake indicating that nano-rock phosphate (RP) coating reduced the availability of N from urea in both the soils. Also, *P* uptake by maize increased with the increase in loading the nano-rock phosphate in both the soils. The results thus indicate that *P* supplied through nano-rock phosphate could be effectively utilized by plant. Hence, coating urea with nano-rock phosphate may be the most economic option for choosing a slow release urea fertilizer.

Trends in N_2O Emission During Incubation Period

The periodic trend in emission of N_2O from different coated urea fertilizers during 33 days period was depicted in Fig. 1. From the results, it was observed that on the first day after application of coated fertilizers to soil showed the highest N_2O emission (0.199 μg N_2O mg^{-1} N) except POR-coated urea (0.102 μg N_2O mg^{-1} N). Thereafter the emission of N_2O decreased up to third day in all the treatments. From third day onwards there was fluctuation in N_2O emission up to 14 days in all the

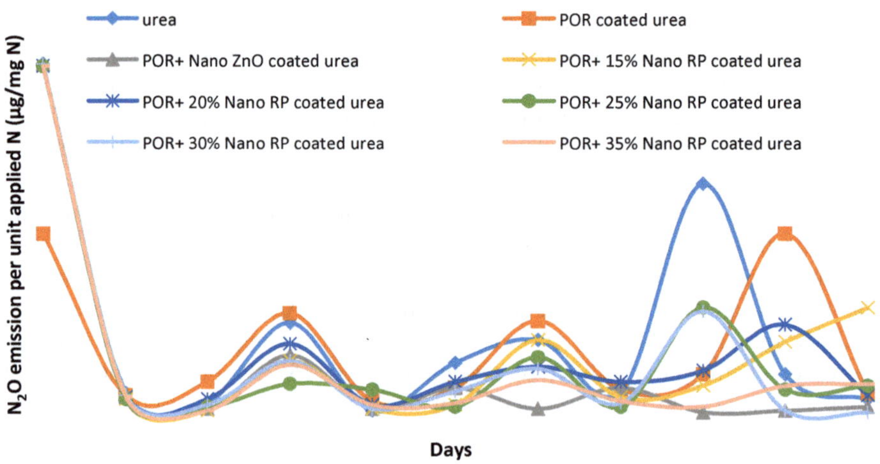

Fig. 1 Trend in emission of N_2O (μg/mg N) from nano-ZnO and nano-rock phosphate-coated urea

treatments, the N_2O level did not cross the 0.06 μg N_2O mg^{-1} N during this period. Thereafter drastic increase in N_2O emission was observed up to 18 days particularly in urea, POR-coated urea, and 15 and 20% nano-RP coated urea applied soils and then decreases gradually but nano-ZnO-coated urea did not show any changes in N_2O emission during this period. Generally when urea was applied to soil, the microbial process like urea hydrolysis and nitrification yield NO_3 that acts as substrate for de-nitrification process for production of N_2O by microbes. Antimicrobial properties of nano-ZnO particles (Espitia et al. 2012; Singh and Nanda 2013) might have influenced the microbial activity that resulted in reduction of urea hydrolysis and nitrification process that resulted in reduction of NO_3 production and subsequently reduced the N_2O emission. Though POR-coated urea initially (up to 3 days) showed less emission of N_2O, in later stages N_2O emission was higher and it was at par with 15 and 20% nano-RP-coated urea. Low emission of N_2O in the initial periods of POR-coated urea might be due to acidic nature and hydrophobicity of POR caused the reduction of influence of microbes and water to interact with urea (Kundu et al. 2013).

Cumulative Emission of N_2O from Coated Fertilizers

In comparison with normal prilled urea, POR (5%)-coated urea application to soil reduced the N_2O emission to the extent of 20.26% during a period of 33 days (Table 5). It might be due to slow release of N from coated urea because of physical barrier by coating and inhibition of urease activity through its antibacterial properties (Trapp and Croteau 2001; Ghosh et al. 2002; Kundu et al. 2013). Further

Table 5 Emission of N_2O from different coated urea fertilizers after 33 days

Coated urea fertilizers	Concentration of N_2O in the head space (μl/L)	Total N_2O in the head space (μl)	Total mass of N_2O in the head space (μg)	Total mass of N_2O-N in the head space (μg)	N_2O-N emission per unit applied N (μg/mg N)	Reduction in N_2O emission (%)
F_1	22.27	2.6733	4.6887	2.9834	0.3242	–
F_2	17.76	2.1318	3.7389	2.3790	0.2585	20.26
F_3	12.26	1.4718	2.5813	1.6424	0.1785	44.95
F_4	17.81	2.1383	3.7503	2.3863	0.2593	20.01
F_5	17.60	2.1125	3.7051	2.3575	0.2562	20.97
F_6	15.95	1.9138	3.3566	2.1358	0.2321	28.41
F_7	15.12	1.8152	3.1836	2.0257	0.2201	32.10
F_8	13.32	1.5993	2.8050	1.7848	0.1940	40.17
CD ($P = 0.05$)	*1.42*	*0.190*	*0.333*	*0.210*		

inclusion of 2% nano-ZnO to POR coating decreased the N_2O emission to the extent up to 44.95% and it was observed to be the highest value in current investigation. This was also the result of antimicrobial properties of nano-ZnO (Espitia et al. 2012; Singh and Nanda 2013). Also it was observed that addition of 15 and 20% nano-RP did not show any drastic reduction in the N_2O emission as compared to POR-coated urea. The reduction of about 20% N_2O emission in these materials could be due to presence of POR coating but not due to nano-RP in the material. Whereas increasing nano-RP to 35% in coating showed significant effect in the reduction of N_2O emission (40.17%). The reduction in the N_2O emission in 35% nano-RP-coated urea could be of enlarged coating that acted as physical barrier that reduced the release and availability of N for microbes to act upon. The cumulative emission of N_2O from nano-ZnO and nano-rock phosphate coated urea during 33 days incubation period was presented in Fig. 2. Among the treatments nano-ZnO-coated urea showed the least N_2O emission of 0.28 μg N_2O mg^{-1} N followed by 35% Nano-RP-coated urea (0.30 μg N_2O mg^{-1} N). The highest N_2O emission was observed in prilled urea (0.51 μg N_2O mg^{-1} N) followed by 15% nano-RP coated urea (0.407 μg N_2O mg^{-1} N) and POR (0.406 μg N_2O mg^{-1} N).

Conclusion

From the study, it was concluded that action of physical barrier through coating, change in microsite pH and antimicrobial properties of coated materials (POR, nano-ZnO and nano-RP) led to slow release and reduced availability of N to microbes, thereby, reduced the potential loss of N as N_2O through de-nitrification and increases nutrient use efficiency. Hence, fortification of urea with 2% nano-ZnO

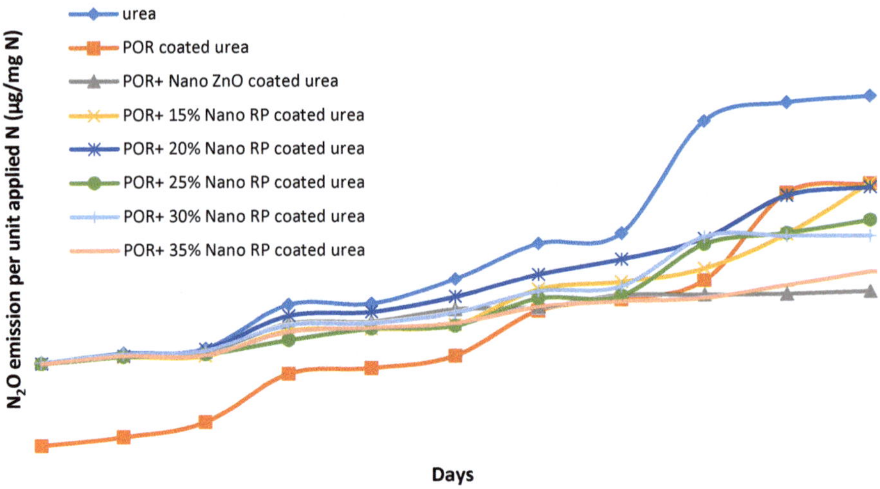

Fig. 2 Cumulative emission of N_2O (μg/mg N) from nano-ZnO and nano-rock phosphate-coated urea

and 35% nano-rock phosphate particles using POR as binding agent could be used as a potential means to reduce N_2O emission from fertilizer N applied to the soil and for increased crop yield.

Acknowledgements This work is the part of NAIP project entitled "Nanotechnology for enhanced utilization of native Phosphorus by plants and higher moisture retention in arid soils." We acknowledge NAIP (ICAR) for the support and funding.

References

Baldwin D, Loeblich V, Lawrence R (1958) Acidic Composition of Oleoresins and Rosins. Ind Eng Chem Chem Eng Data Ser 3:342–346

Espitia P, Soares NDF, Coimbra JSDR, Andrade NJ, Cruz RS, Medeiros EAA (2012) Zinc oxide nanoparticles: synthesis, antimicrobial activity and food packaging applications. Food Bioprocess Technol 5:1447–1464

Galloway JN, Cowling EB (2002) Reactive nitrogen and the world: 200 years of change. AMBIO J Hum Environ 31:64–71

Ghosh BN, Chowdhury H, Kundu S, Gupta HS (2002) Effect of pine needle (*Pinus roxburghii*) and chinaberry (*Melia azaderach*) seed extracts on urea hydrolysis rate in a sandy soil. J Indian Soc Soil Sci 50:309–311

Junejo N, Khanif MY, Dharejo KA, Abdul-Hamid Hazandy, Abdu Arifin (2011) A field evaluation of coated urea with biodegradable materials and selected urease inhibitors. Afr J Biotechnol 10:19729–19736

Kundu S, Adhikari T, Vassanda Coumar M, Rajendiran S, Bhattacharyya R, Saha JK, Biswas AK, Subba Rao A (2013) Pine Oleoresin: a potential urease inhibitor and coating material for slow-release urea. Curr Sci 104:1068–1070

Ladha JK, Pathak H, Krupnik TJ, Six J, Van Kessel C (2005) Efficiency of fertilizer nitrogen in cereal production: restropect and prospects. Adv Agron 87:85–176
Lloyd WD, Hedrick GW (1965) Levopimaric acid. Org Synth 45:1–64
Pathak H, Aggrawal PK (2012) Low carbon technologies for agriculture: a study on rice and wheat systems in the indo-gangetic plains. Indian Agricultural Research Institute, New Delhi, p xvii + 78
Pathak H, Gupta PK, Bhatia A, Sharma C, Kalra N, Mitra AP (2004) Nitrous oxide emission from soil-plant system. In: Mitra AP, Sharma S, Bhattacharya S, Garg A, Devotta S, Sen K (eds) Climate change and India, uncertainity reduction in green house gas inventory estimates, pp 244–257. Universities Press (India) Pvt. Ltd., Hyderabad
Prasad R (1998) Fertilizer urea, food security, health and environment. Curr Sci 75:677–683
Purkayastha TJ, Chhonkar PK (2006) Evaluation of ammonia volatilization from compacted urea-micronutrient fertilizers applied in rice (*Oryza sativa* L.) grown on a typic haplustept. J Indian Soc Soil Sci 54:1–10
Singh P, Nanda A (2013) Antimicrobial and antifungal potential of zinc oxide nanoparticles incomparison to conventional zinc oxide particles. J Chem Pharm Res 5:457–463
Smil V (2001) Enriching the earth: Fritz Haber, Carl Bosch and the transformation of world food production. MIT Press. ISBN 026219449X, 411 pp
Trapp S, Croteau R (2001) Defensive biosynthesis of resin in conifers. Annu Rev Plant Physiol Plant Mol Biol 52:689–724

Gamma Radiation Technology for Hygienization of Municipal Dry Sewage Sludge

Naresh Kumar Garg and Lalit Varshney

Abstract Sewage is the wastewater discharged from domestic premises consisting mainly of human waste. Sewage typically contains more than 99.9% water and about 0.05% solid. The solid part results in the formation of sludge. Largely, sludge is disposed in an unorganized manner resulting in environmental pollution and spread of diseases. The sludge produced carries a heavy microbiological load including pathogens and therefore its disposal has been a challenge to the urban development authorities. Indian cities and towns together are generating an estimated sewage load of 38,254 million liters per day (MLD), out of which 11,787 MLD is treated at sewage treatment plants (STP) with a capacity gap of 26,737 MLD (Ministry of Urban Development 2013). Considering 0.05% solid content, the total potential of sludge generation from the sewage is 19,127 tons per day. High energy ionizing radiation technology has a great potential for hygienizing municipal sewage sludge and makes it safer for use or disposal. In this study, dry sludge hygienization process was studied and compared with wet sludge hygienization. Dry sludge in plastic bags was hygienized in a terminal process using standard fully automatic irradiation plant without manual intervention. Dry sludge containing 75–80% solid irradiated at an average dose of 8 kGy, showed the absence of indicator organism *E. coli*/total coliforms even after 10 months of study period. Heavy metal concentrations in the domestic sewage sludge were observed to be much below the United States Environmental Protection Agency (US EPA) and Ministry of Urban Development, New Delhi (MOU) norms (Ministry of Urban Development 2013; United States Environmental Protection Agency). Inoculation of the hygienized sludge with Rhizobium, Azotobacter, and Phosphate solubilizing bacteria showed 100–1000 times higher growth in comparison to growth in unhygienized sludge. A comparison between dry and wet sludge (4–6% solid) irradiation shows that it is more practical to hygienize dry sludge which is economical, reliable, and scalable to treat 100–500 tons of city dry sludge. Use of

N.K. Garg (✉) · L. Varshney
Radiation Technology Development Division, Bhabha Atomic Research Centre, Mumbai 400085, India
e-mail: nkgarg3@gmail.com

V.P. Singh et al. (eds.), *Energy and Environment*, Water Science and Technology Library 80, https://doi.org/10.1007/978-981-10-5798-4_10

radiation technology for sludge hygienization can significantly contribute to "Clean India Mission". The technology is available with the department.

Keywords Sewage sludge · Dry sludge · Wet sludge · Gamma radiation Radiation hygienization · Pathogens · Microorganisms · Radiation dose Over dose ratio

Introduction

STP-generated dry sludge contains heavy load of bacterial counts including pathogens which could vary between 10^5 and 10^9 per gram. Sludge carries other life forms like worms, ova, viruses, helminthes, weeds, etc. It contains toxic heavy metals and organic pollutants like pesticides, polyaromatic hydrocarbons, drugs, and other persistent pollutants. Sludge is also a rich source of many macro-(Nitrogen, Phosphorous, Potassium), and micro-nutrients (Zinc, Iron, Copper, Manganese) and organic carbon essential for soil. Hygienized sludge can be gainfully used for agriculture, road side plantation, forestry, city home plants, etc. Ionizing radiation like gamma rays from cobalt-60 and electron beam from accelerator can be employed to deliver sufficient radiation dose to sludge to inactivate all life forms in the sludge and hygienize it (International Atomic Energy Agency 2002). The sewage sludge can be treated by two methods: (1) Wet sludge Irradiation (4–10% solid content, 96–90% water) and (2) Dry Sludge Irradiation (~75–80% solid, 25–20% water). The end product of both the process is dry solid cake.

Experimental

Plastic bags were filled with dry sludge (25 kg each) from drying beds of the STP and irradiated in a standard irradiator with conveyor system having about 700 kCi Cobalt-60 source. Ceric-cerous sulfate dosimeters and microbiological dosimeters were placed at minimum and maximum dose positions inside the bags and were sealed. Electrochemical method was used to evaluate absorbed radiation dose. Absorbed dose measurement and microbiological studies were carried out as per standard procedures (Gautam et al. 2005). The process of wet sludge hygienization is described elsewhere (Gautam et al. 2005). Reported values are an average of three readings. Microbiological indicator disks containing 1×10^6 spores of *Bacillus pumilus* ATCC14884 and Ceric-Cerous Sulphate Dosimeters were obtained from Board of Radiation and Isotope Technology (BRIT).

Results and Discussion

The quality of sludge (dry, semi-dry, or wet sludge) is determined by three factors namely (a) the presence of pollutants (arsenic, cadmium, chromium, copper, lead, mercury, nickel, selenium, and zinc. (b) The presence of pathogens (e.g., bacteria, viruses, parasites) and (c) The sewage sludge attractiveness to vectors, e.g., rodents, flies, mosquitoes, birds, etc.(United States Environmental Protection Agency). The process followed by STPs and determination of heavy metals could help in checking the suitability of sludge for use. In wet sludge irradiation, at 3–5 kGy dose, a larger portion of energy is consumed by water (96%) and output of dry sludge is only 4%, as well as the irradiator would require large quantity of Cobalt-60 and thus making economics unfavorable. Irradiation of dry sludge containing about 80% solid at an average dose of 8–10 kGy gave 20 times more output and lower radiation source requirement per kg of hygienized sludge produced. The results of microbiological investigations on irradiated dry sludge are given in Table 1. US EPA recommends use of treated sludge within 8 h of pathogen reduction process to overcome regrowth issue (United States Environmental Protection Agency). Hygienized dry sludge in plastic bags did not show increase in *E. coli*/total coliforms even after 10 months of study period. Reduction by about four log cycles of high radiation-resistant *B. pumilus* in the microbiological indicator (Table 1) demonstrates efficacy and reliability of radiation treatment for hygienization of dry sludge.

The results of inoculation studies showed 100–1000 times more growth of useful bacteria in the hygienized sludge in comparison to unhygienized sludge. The US/EPA requirements of heavy metals and microorganisms are given in Table 2. The heavy metal concentration in domestic sewage (Collected at SHRI facility, Vadodara) was observed to be much lower than the norms and therefore can be applied to land. Irradiation degrades toxic chemicals and does not alter the heavy metal concentration in the sludge (International Atomic Energy Agency 2002).

Table 1 Microbiological investigations on dry sludge

S. No.	Sample no.	Average dose (kGy)	Total [a]coliform/g	*Bacillus pumilus* spores after irradiation	Total coliform/g after 10 months
1	Control		11.5 × 106	1 × 106	8.7 × 103
2	A1 (minimum)[b]	6.85	NIL	0.96 × 102	NIL
3	A2 (maximum)[b]	9.03	NIL	0.30 × 102	NIL
4	B1 (minimum)	7.07	NIL	0.42 × 102	NIL
5	B2 (maximum)	9.15	NIL	0.16 × 102	NIL
6	C1 (minimum)	6.81	NIL	0.85 × 102	NIL
7	C2 (maximum)	9.04	NIL	0.10 × 102	NIL

[a]US, EPA recommends less than 1000 CFU/MPN per gram at the time of disposal
[b]Minimum and maximum dose positions in the bag containing dry sludge

Table 2 Toxic metal limits[a] in sludge

Pollutant	Ceiling[a] concentration limits for all biosolids applied to lands (mg/kg)	Typical Conc. in SHRI[b] sludge dry basis (mg/kg)	Annual pollutant loading rate limits for APLR biosolids (kg/hectare per 365 day period)
Aresenic	75	–	20
Cadmium	85	<10	1.9
Chromium	3000	–	150
Copper	4300	190	75
Lead	840	125	15
Mercury	57	Not detected	0.85
Molybdenum	75	–	–
Nickel	420	65	21
Selenium	100	–	5.0
Zinc	7500	620	140
Applies to	All biosolids that are land applied	–	Bagged biosolids

[a]US EPA 40 CFR 503
[b]SHRI is a wet sludge hygienization research irradiator at Vadodara, India

On meeting the norms, the hygienized sludge can be safely applied on land applications. The radiation technology for dry sludge hygienization is available with the department.

Acknowledgements The authors acknowledge the support and encouragement given by Dr. Sekhar Basu, Director BARC, Dr. K.L. Ramakumar, Director RC&IG and Dr. A.K. Kohli, CE BRIT for developing Radiation Technologies for societal benefits.

References

Gautam S, Shah MR, Sabharwal S, Sharma A (2005) Gamma irradiation of municipal sludge for safe disposal and agricultural use. Water Environ Res 77(5):472–478

Irradiated sewage sludge for application to cropland, IAEA-TECDOC–1317, International Atomic Energy Agency (October 2002) pp 1–3

Manual on sewerage and sewage treatment system, Ministry of Urban Development, Part A, Page nos.- November 2013, New Delhi. (http://moud.gov.in)

United States Environmental Protection Agency (US, EPA), 40 CFR Part 503, Biosolid Rules

Part IV
Environment: Social Environment

Participatory Approach for Corporate Social Responsibility Plan in India

Rajesh Puranik

Abstract The ministry of environment, forest and climate change (MoEF&CC) has notified the environmental impact assessment (EIA) Notification, 2006 under the provisions of the Environment (Protection) Act, 1986, which regulates development and expansion/modernization of 39 sectors/activities listed in the Schedule to the EIA Notification, 2006. There are two category of the projects viz. Category 'A' projects are handled at the level of MoEF&CC and the Category 'B' projects are handled by the respective State Environment Impact Assessment Authority (SEIAA) following the procedure prescribed under the EIA Notification, 2006. Developmental projects in various sectors need environmental clearance either SEIAA or MOEF&CC (MoEF) depending upon their categories and scale of development. A project proponent, any person desiring to establish his industrial unit or project is required to apply for environmental clearance (EC) before establishing project or industry for getting the EC. Standard procedures of development of environmental impact assessment (EIA) Report are prescribed by the authorities and are available in public domain. Project Proponents approach EIA consultants to prepare their EIA reports as per the norms and standard procedures. EIA consultants are the persons or group of persons having experience and specialization in various environmental aspects and are conversant with various policies applicable at the state or national level. The government of India, MoEF&CC has adopted various measures to improve the quality of EIA reports prepared by the consultants. Recently, grading of consultants and their accreditation has been enforced through quality council of India (QCI) under the scheme National Accreditation Board for Education and Training (NABET). Only NABET approved consultants and laboratories are permitted to prepare EIA reports and make presentations on behalf of Project Proponent to state or MoEF&CC authorities. Various studies of environmental parameters are undertaken by the expert EIA consultant organization, results are tested in approved laboratories and reports are generated for EC. MoEF&CC has also prescribed the preparation of social impact

R. Puranik (✉)
Social Science Faculty, WALMI, Aurangabad, Maharashtra, India
e-mail: rajeshpuranik@gmail.com

V.P. Singh et al. (eds.), *Energy and Environment*, Water Science and Technology Library 80, https://doi.org/10.1007/978-981-10-5798-4_11

assessment reports (SIA), socio-economic survey reports (SES), ecology and biodiversity (EB) reports, rehabilitation and resettlement plans (R&R) and corporate social responsibility plans (CSR) to be a part of EIA reports. Standard terms of reference (TOR) for EIA/EMP report for projects/activities requiring environment clearance has been issued by MoEF&CC in April 2015. It contains a general condition for CSR that "along with details of villages and specific budgetary provisions (capital and recurring) for specific activities over the life of the project should be given." This paper deals with methodology of preparation of CSR plans as per the felt needs of project affected people. Classically, CSR is considered as a responsibility of project proponent to compensate losses to the environment and people that are directly or indirectly affected due to project. This paper gives insight on developing the CSR plan as per felt need of people in core as well as peripheral zone of the project. Existing approach of preparation of CSR plan can be benefited by the proposed methodology and project proponents can deliver their social responsibilities keeping people and environment in focus. The participatory approach of involving people and various techniques of participatory rural appraisal (PRA) and logical framework analysis (LFA) can be used to collect and synthesize the information for developing CSR plans; in fact, people themselves develop the plan.

Keywords SIA · SES · R&R · EB · CSR · PRA · LFA · EIA
EMP · Plan · Methods

Introduction

Growth and development of any region often depend on introducing urban and industrial activities in a planned way. Baseline data on socio-economic parameters such as demography, infrastructure, economic resource base, health status, cultural aspect and aesthetic attributes are generated using information available with the Government agencies, census books, statistical abstracts and health agencies. The opinions of local people about the proposed action are also obtained during socio-economic survey carried out in core and buffer zones villages selected close to the proposed project site.

The socio-economic study is generally set in the Core Zone (the area of project establishment/plant erection) as well as Buffer Zone (the area falling within the radius of 10 km of the core zone). The study assesses and evaluates broad economic resource base and socio-economic profile of the villages with an implicit purpose of assessing socio-economic impact of proposed project and its policies addressing various issues including corporate social responsibility (CSR) of project proponent. Hence while preparing CSR plan it is equally important to carry out socio-economic baseline studies or refer to such studies if already carried out for the area under project.

It may be asked why a corporate person should take into consideration the Social Responsibility? The answer to this is very simple and logical. In fact, it has been well accepted that development sector is epicentre of our sustainable development and people-centred policies are taking root not only in public enterprises but also in private sectors. Investments made to improve the service delivery and upgrade the value chain in India in last few decades are witnessing the unprecedented change in outlook of people and greater awareness among them.

Recently in the year 2013, a mandate for corporate social responsibility has been formally introduced in the Companies Act by Government of India. Response to the initiatives suggested for CSR by government will be on test of time for various public, private and multinational industries/organizations. The practice of CSR is not new to companies in India. However, what New Company Act 2013 does is to bring more companies into the fold. Also, it is likely that the total CSR spends will increase. What is clear to many companies is that if this increased spending is to achieve results on the ground—which is the intent of the Act—then it needs to be done strategically, systematically and thoughtfully (Mittal 2013). Provisions in the new companies act call for structured and participatory engagement of companies with communities especially for planning and designing of social and environmental impact mitigation measures through their environment impact assessment (EIA) reports and environment management plans (EMP) comprising of need base CSR plan and socio-economic survey (SES) baseline studies. Now every industry/enterprise is required to have their own CSR policy, strategy and programmes with some dedicated manpower who are responsible for implementation of CSR plans of the said unit. Few organizations are practicing CSR for many years and discharging their responsibilities towards the society. However, it was not common and popular among many other organizations to adopt social responsibilities on their own before introduction of new act and its provisions. Hence it is an opportunity to build a society which provides equal access to harness the resources and also negate the disparities and promotes collective responsibility. This paper discusses a participatory process of ensuring utilization of CSR funds to fulfill the felt needs of the communities, which are identified and prioritized by community themselves.

Socio-economic Baseline Data for CSR Plan

Identification of needs of villagers to develop Corporate Social Responsibility plan is the first step towards people's involvement in CSR activities. Objectives of socio-economic baseline study should include:

- Collection of socio-economic and demographic details of areas by adopting the participatory approaches.
- Identification of felt needs and income generation activities preferred by villagers and develop business plans for selected options.

- Identification of impacts of the project on community as a whole.
- Development of Corporate Social Responsibility Plan to address the felt needs as expressed by communities.

Participatory Survey

Basic information is always required for planning the CSR and community development programs which depend on the demographic parameters of the project area. This includes total population, their livelihood, total number of families, family size wise classification of households, age composition, etc. Similarly, and equally important is the data related to the occupational pattern of the population and relative size of each occupational group.

Precise information on availability or necessity of various infrastructure related to health, education, transport, drinking water, electrification and other civic amenities is also necessary for planning. A participatory survey by involving communities is the key to develop a sound foundation for CSR plans.

Participatory Approach

Participatory approaches are now commonly adopted by the practitioners, social and economical experts in the field. Most popular among these are the participatory rural appraisal (PRA) tools, rapid rural appraisal (RRA) tools, logical framework analysis (LFA), etc. A basic concept of participatory survey is the community assessment of available resources and analysis of vision for their own improvement. The funds available with the corporate enterprises can be utilized for CSR activities.

Team Building

Team building for collection of facts from the field is the most important activity. Every member of the team must have a clear understanding of purpose of the activity and shall be able to use the participatory approaches effectively to collect information from the field. Similarly, selection of supervisors and field workers/enumerators/animators especially from social work background is of equal importance. It is imperative to organize one training-cum-workshop before actual field survey is undertaken.

Preparation of FGD Guides: Separate instrument in the form of FGD Guide is required in local vernacular language in the light of the objectives of the activity. The questions in the guide need to be worded clearly in a simple manner in order to make it easy for the respondents to understand and reply without any difficulty and express their opinion freely and frankly. Direct observation of the site by the surveyors is the important part of the activity. A check list to record the facts in support of the respondents' replies is also required wherever possible. These instruments can then be administered in the field for collecting the facts and felt needs of the villagers.

Data collection: For collection of data, graduate and post graduate youth with qualifications in social work may be preferred as enumerators. These enumerators with previous experience of conducting similar kind of task can be instrumental in collecting the error-free data. Enumerators having cognizance of local cultural value system may be preferred during selection process. Selected enumerators are then required to be made aware of the specific purpose of the activity and trained to seek required information.

Focused Group Discussion: Focused Group discussion is one of the effective tools of PRA for gathering the information on specific issues which need general consensus of people at large. FGD can be conducted among different stakeholders with the help of the FGD Guides at some at common places in the villages where respondents had easy access and could participate freely in the discussions. Exclusive FGD sessions with the vulnerable groups, especially the women, may be necessary to have insight into their specific needs.

Participatory Rural Appraisal (PRA): PRA is also conducted in the villages to collect the factual information by involving the community. Different tools of PRA like transect walk and reconnaissance survey, etc. are used to collect the information.

Infrastructure Survey: Survey of the existing infrastructure in villages is useful in improving the infrastructure as well as planning for its development. Project team collects all the relevant information in a participatory way during various meetings, discussions with villagers, local leaders, etc. This comprises of collection of information on all items of basic facilities like roads, street lights, water resources, pasture lands, open areas, forestry resources, cremation/burial grounds, extent of agricultural land, irrigation, educational facilities, health services, veterinary healthcare services, business units, service industries, panchayat and infrastructure that exist in the villages. The data on availability of and access to infrastructural facilities is also compiled from census 2011 or the latest Census Data and adequate survey in selected villages.

Common Property Resource Survey: Transect walk is used as a tool whereby animators take an observational walk with villagers during which due attention is given to people, resources, environmental features, physical setup, etc. The community members provide useful information and clarify the doubts. The output of transect walk is the hands-on experience in a community to understand the village setup and identification of common property resources that exist within the villages.

Village information: Village information schedule is used for collection of village information. The following information is collected.

A. General information of village, its geographical location etc.
B. Demography
1. Population as per Ward/Tola
2. Marital status
3. Migration status
4. Reasons for migration
5. Population distribution
C. Social structure
6. Size of family
7. Housing
8. Social composition
9. Educational status
D. Economical status
10. Occupation
11. Cropping pattern
12. Milk, eggs and fish production
13. Land holding
14. Labour class status
15. Labour rate (existing Rs. per day)
E. Status of villages
16. Ground water level
17. Surface water availability
18. Source of water
19. Biomass cover
20. Energy
21. Community—groups
22. Agricultural implements
23. Parental businesses

Socio-economic Information: A pre-designed questionnaire is generally used for collecting socio-economic information of the village households. Following important information is included in the schedule:

- Village infrastructures and its status
- Village facilities and its status
- Professional training need of villagers
- Socio-economic profile of village households

Capital investment activities

S. No.	Work	S. No.	Work
A	Modular and non modular businesses	VIII	Sector: education
I	Sector: modern agriculture equipments and tools	IX	Sector: entrepreneurship development

(continued)

(continued)

S. No.	Work	S. No.	Work
II	Sector: animal husbandry and dairy works	X	Sector: water related infrastructure development
III	Sector: social forestry	XI	Sector: community development and community information center
IV	Sector: strengthen agriculture	XII	Sector: community-related infrastructure development
V	Sector: health	XIII	Sector: education related infrastructure development
VI	Sector: drinking water	XIV	Sector: solid/liquid waste management
VII	Sector: community infrastructure	XV	Sector: school sanitation hygiene education

Recurring cost investment activities

S. No.	Particular	S. No.	Particular
I	Strengthen off farm and non-farm occupations (local unemployed youths)	VI	Social awareness campaigns
II	CBOFCB: community based organizations for capacity building	VII	Health awareness to improve health
III	Strengthen irrigation and agriculture facilities in villages—farmer's training	VIII	Minor forest produce
IV	Capacity building of subsidiary enterprises	IX	Maintenance and repair
V	Improve functional literacy	X	Strengthen extension services

FGD guide/observation format

Areas	Observations and photographs
Health	
Public health	
Veterinary clinic	
Education	
Schools/anganwadis	
Anganwadi kitchen shed	
Livelihood	
Handloom/weaving, etc.	
Agriculture and allied	
Forest/NTFP/local species	
SHGs	
Animal husbandry	
Dairy	
Sericulture	

(continued)

(continued)

Areas	Observations and photographs
Social/agro forestry	
Horticulture	
Water resources, irrigation source (river/well/tube well)	
Enterprise development	
Local skills	
Soil and water conservation	
Infrastructure	
Road/connectivity	
Electricity/solar power	
Drinking water/dug-well/tap water	
Sanitation/water SLWM (solid/liquid waste management)	
Playground/public park, etc.	
General utility, community infrastructure/gotul (community hall constructed for tribal villages)/waiting room, etc.	
Markets/haats/bazaars/pashu bazaars, etc.	
Water bodies (lake, pond etc.)	
Entrance gate	
Computers/internet, etc.	
Haudis (drinking water arrangement for animal, cattle)	

Impact of CSR Activities: The following is an illustrative list of possible impacts:

Social/Cultural

- community cohesion
- integration of social support systems
- Development of women's economic activities
- Restoration of time-honoured sacred places of worship
- Restoration of archeological sites and other cultural heritage property

Economic

- Benefit to agricultural lands, tress, wells
- Benefit to dwellings and other farm buildings
- Benefit to access to common property resources
- Benefit to shops, commercial buildings
- Benefit to businesses/jobs
- Overall Benefit to in income due to above developments

Improvement to Public Infrastructure and services

- Government office buildings
- School buildings

- Hospitals
- Roads
- Street lighting

Other Impacts

Impact on livelihood: Impact of the project is exemplified by people becoming landless, removal of their main occupational foundation and affecting their commercial activities and livelihoods. CSR activities restore people's economic and commercial activities by improving their skills and providing them opportunities for doing modular and non-modular businesses.
Creation of job opportunities: Joblessness or loss of employment and wages deprive landless labourers, service workers, artisans, and small business owners of their sources of income due to project are restored with CSR activities.
Community Development: Homelessness or loss of housing and shelter threatens to become chronic for the most vulnerable. In a broader cultural sense, it also affects communities' cultural space and identity. Community development activities of CSR help communities in restoring their cultural space and culture.
Improvement in upliftment of project affected persons: Marginalization is one of the predicted impacts of project which occurs when families lose economic power. Middle-income farm households become small farmers; small shopkeepers and craftsmen are downsized and slip below poverty thresholds. Economic marginalization is often accompanied by social and psychological marginalization and manifests itself in a downward mobility in social status, displaced persons' loss of confidence in society and in themselves, a feeling of injustice and increased vulnerability. CSR activities combined with environment mitigation plans in project affected area provide opportunity to rural farm households, artisans, small shopkeepers, craftsmen to improve their sources of income and economic status, and thereby ensuring them financial stability and restoring their confidence.
Food security: Regular income, increased income source and economic stability overcome the food insecurity.
Reduction in Morbidity and Mortality: There is increased accessibility to health facilities due to infrastructure development under the CSR activities which reflects in good health, minimizing malnutrition, reducing stress and psychological traumas. Improvement in village facilities provides safe water supply and waste disposal which further reduces possibilities of proliferation of infectious disease.
Increased Access to Common Property: Increased access to commonly owned assets (forestlands, water bodies, grazing lands and so on) is possible due to various developmental initiatives and it compensates the losses in the long run.

Steps in Conducting CSR

Step 1: **Define the CSR Area**
The first step is to define the *Area of CSR activities*. The CSR team must get a map showing clearly demarcated area that will be brought under CSR activities. Field visit to the area needs to be undertaken to have a better understanding of the geographic limits of the area and the people living there.

Step 2: **Identify Information/Data Requirements and their Sources**
Review the existing data on various developmental activities being undertaken by public and private agencies. This may provide basis for planning appropriate CSR initiative and converge the activities and funds with other developmental initiatives to get synergy effect. This review will also help identify the need for collection of additional primary data through surveys and participatory methods.

Step 3: **Involve Stakeholders in planning CSR**
Share information of CSR and consult with all stakeholders. Stakeholders are people, groups or institutions which are living in the identified area proposed for CSR intervention. This sharing of information helps to develop and implement an effective public involvement plan. The first step in developing plans for public consultation and participation is to identify stakeholders who need to be involved in the consultative processes. The basic questions to consider in identifying stakeholders include:

- Who will be directly or indirectly and positively and negatively affected?
- Who are the most vulnerable groups, who need CSR support?
- Who might have an interest or feel that they are affected?
- Who supports or opposes the CSR activities or the changes that the project produces?
- Who are opposing the CSR initiatives and why?
- Whose cooperation, expertise or influence would be helpful to the success of the CSR activities?

Step 4: **Stakeholders Consultation in the Field**
Stakeholder's consultation is conducted in field at common place accessible to all stakeholders. It is important to confirm their understanding of key issues. The local knowledge can be invaluable in finding alternatives that help to restore or cover developmental interventions to avoid or at least reduce the magnitude and severity of adverse impacts. This is done during initial assessment and can be used for further in-depth inquiries for need-based CSR interventions.

Step 5: **Prepare a Socio-economic Profile of Baseline Condition**
To assess the extent of impact of CSR activities, it is necessary to assess the socio-economic conditions in the project area. This assessment generally involves conducting a socio-economic survey and a broad-based consultation with all affected groups.

The socio-economic profiling is done while social impact assessment is conducted, and the same can be used for baseline socio-economic analysis. However, it should not be restricted to adversely affected population only. The survey should include those who benefit from the employment and other economic opportunities generated by the project.

CSR Methods and Tools

CSR planning should be carried out before the start of the project. The basic objectives of this study are to provide

- Interventions to meet the people's felt need
- Interventions to reduce the potential impacts of the project, its magnitude, distribution, and their duration;
- Information on who will be benefited, individual or group, positively or negatively affected
- Information on perceptions of the people about developmental initiatives
- Information on potential mitigation measures to minimize the impact
- Information on institutional capacity to implement mitigation measures

Examples of Questions to Be Addressed in CSR Planning

- Who are the key stakeholders? What do they already know about the proposed project, and the measures including CSR activities being contemplated to mitigate its negative impact?
- What are their interests? Are the objectives of the CSR are in line with their needs, interests and capacities?
- What will be the impact of CSR on various stakeholders, and particularly on women and vulnerable groups?
- What social factors affect the ability of stakeholders to participate or benefit from CSR (gender, caste, ethnicity, or income level)?
- What institutional arrangements are needed for implementing CSR interventions?

Sources of Information

The CSR Plan Requires Both Primary and Secondary Data

(a) Primary Source: The existing data from secondary sources cannot, however, be a substitute for project-specific surveys. In addition, SIA derives much more relevant information directly from surveys of various kinds including socio-economic survey, and FGD and consultation with the affected people
(b) Secondary Source: Such sources of data include:

- Government census data
- Land records, including records of land transactions
- District gazetteers
- Other administrative records (such as NSS)
- Documents from non-governmental organizations

Data Collection Methods for Preparing CSR Plan

Qualitative Methods

(a) *Key Informant Interview*: A set of questions on identifying the social development activities or environmental conservation activities which can be covered under CSR are incorporated in the questionnaire. Such questionnaire is useful in collecting the responses of people in structured manner. The design of the questionnaire is rather important. It should focus on key issues, yet be simple and in the local language. Persons selected to conduct the interviews should be properly briefed and trained to get the questionnaires completed. The team conducting the interviews should include female members, to collect information on women issues especially in rural areas and among communities where there are restrictions on their movements.
(b) *Focused Group Discussions (FGDs)*: In FGDs, one or more informants guide a group discussion and encourage the group members to discuss the topic among themselves. The group usually has six to ten participants for better interactive session. However, more participants are also encouraged to join the discussion to get a proper representation from all social sections. The informant usually follows an interview guide but interferes minimally in the discussion.
(c) *Participatory Rural Appraisal*: This low-cost method is based on in-depth interactions with rural people and involving them in the process of data generation. There are several PRA tools which are used to generate the information which later is used for planning purpose.

Quantitative Methods

(a) *Common property assets survey*: Common property assets survey is conducted by using questionnaire wherein detailed information of assets available in the village are explored along with its year of origin and present status. This survey helps in understanding the new interventions or management and maintenance requirement of already created assets. The common property survey is expected to provide answers to questions such as the following:

- Where is the property developed?
- Who is the current owner?
- What is present status?
- What is the present use?
- In addition, a comprehensive list of common property is prepared which includes:
- Connectivity: Approach roads, roads, bridges, power lines, and water and sewage lines.
- Pastures, ponds and forests including craft materials, biomass for domestic energy.
- Public structures: These include schools, clinics, places of worship, bathing and washing places, community centres, lampposts, playgrounds, wells and bus stops
- Cultural property: Cultural property includes archeological sites, monuments and burial grounds, places of historical or religious importance.

Reference

Mittal RB (2013) Message from the Chairman, CII Development Initiative Council, Vice Chairman and Managing Director,Bharti Enterprises, published in Handbook on Corporate Social Responsibility in India, www.pwc.in, PricewaterhouseCoopers Private Limited, NJ 123-Nov 2013 CSRHandbook.indd

Part V
Energy: Solar Energy

Solar-Powered Cold Storage System for Horticultural Crops

P.L. Singh, P.C. Jena, S.K. Giri, Baba Saheb Gholap and Omkar Singh Kushwah

Abstract A solar photovoltaic-powered cold storage system has been developed for storage of fresh horticultural produce (6–7 tonne capacity) at Central Institute of Agricultural Engineering, Bhopal. Temperature and relative humidity controllers are fitted in the cold storage chamber to maintain desired room temperature (5–15 °C) and relative humidity (65–95%) for storage of horticultural produce. The battery backup was used for storage of solar power for operation of the cold storage system during night and cloudy weather. The shelf-life of fresh unripe mature mango (Dashari) was increased by 15 days in the cold storage as compared to 4 days at ambient storage during June month. Energy output from the solar power plant ranged from 67 to 110 kWh/day on sunny days in different months, which was sufficient to operate the cold storage unit.

Introduction

India is the second largest producer of horticultural commodities in the world with 88.977 million MT of fruits and 162.887 million MT of vegetables during the year 2013–14 (Indian Horticultural Data base 2014). A large variety of fruits are grown in India, of which mainly banana (33.4%), mango (20.7%), citrus (12.5%), papaya (6.3%), guava (4.1%), grape (2.9%), apple (2.8%), sapota (2.0%), pomegranate (1.5%), litchi (0.7%), etc. are the major ones (Fig. 1). Similarly, the major vegetable crops grown in India are potato (25.5%), tomato (11.3%), onion (11.9%), brinjal (8.3%), cabbage (5.5%), cauliflower (5.3%), okra (3.9%), pea (2.4%), etc. (Fig. 2). In the country, per capita availability of fruits and vegetables is quite low because of post-harvest losses due to poor and inadequate storage infrastructure for perishable products. The Associated Chambers of Commerce and Industry of India reported that about 30% of vegetables and fruit are lost after harvest, worth a total of over

P.L. Singh (✉) · P.C. Jena · S.K. Giri · B.S. Gholap · O.S. Kushwah
Central Institute of Agricultural Engineering, Nabibagh, Berasia Road, Bhopal 462038, MP, India
e-mail: pannalalsingh24@gmail.com

V.P. Singh et al. (eds.), *Energy and Environment*, Water Science and Technology Library 80, https://doi.org/10.1007/978-981-10-5798-4_12

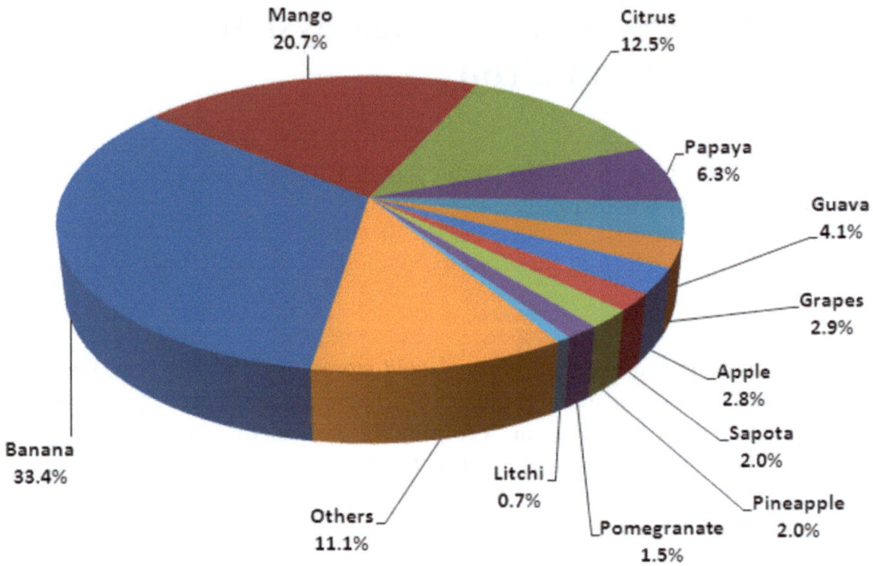

Fig. 1 Production share of major fruits crops in India during 2013–14

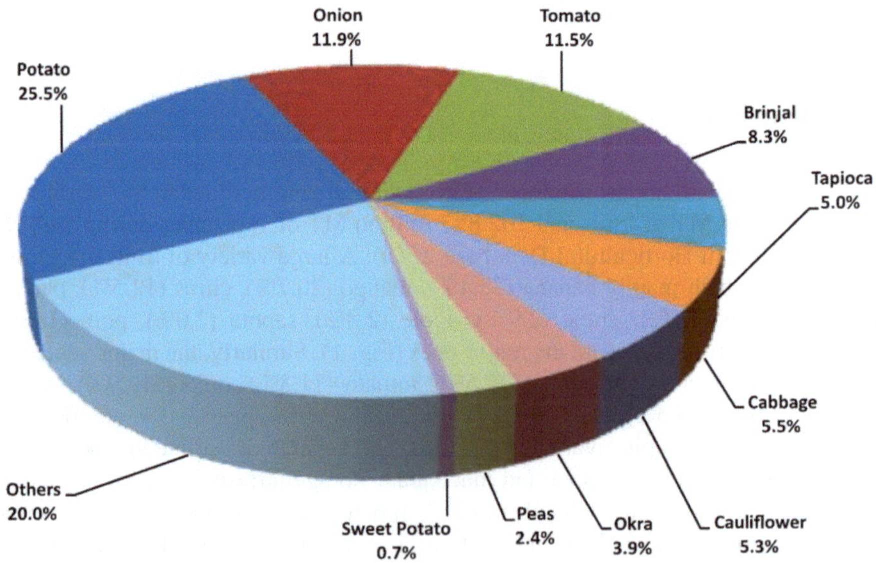

Fig. 2 Production share of major vegetable crops in India during 2013–14

2 trillion INR per year (over 33 billion USD) due to lack of storage and processing facilities and indifferent attitudes towards tackling the problem (ASSOCHAM 2013). Among the states, post-harvest losses are highest in West Bengal (over 136.6 billion INR), followed by Gujarat (114 billion INR), Bihar (107 billion INR) and Uttar Pradesh (103 billion INR). A sizable quantity of produce also deteriorates by the time it reaches the consumer. To enhance shelf life and maintain the quality of perishable products, cold chain arrangement is required if consumption is not meant immediately after harvest. Most of the horticultural produce required cooling temperature between 0 and 15 °C for safe storage and transient purposes. In the absence of a cold storage and related cold chain facilities, the farmers are being forced to sell their produce immediately after harvest which results in glut situations and low price realization. Cold chain infrastructure for fruits and vegetables can improve quality and enhance shelf life of the product. Robust farm-to-retail cold chain solution is required to sustain the growing domestic and export demand. Apart from the large cold storage structures for long-term storage, cooling system is also required for on-farm or in-production catchment for horticultural crops, so that produce get cooled during short-term storage and at the same time it can be loaded in the transportation vehicle in cool condition to maintain the quality and minimize the wastage during transportation.

In conventional cold storage systems used in India, the electrical energy expenses account 28–30% of total expenses (Energetica India 2012). Solar-based refrigeration system is very relevant to India as the running cost reduced by use of solar power. India is blessed with good amount of solar energy in most part of the country throughout the year. The mean annual solar radiation is varied between 4.6 and 6.6 kWh/m^2/day in different parts of the country. Grid power availability in the rural areas is very poor with respect to its supply and quality. The solar power is one of the best solutions for operating small cold storage system in rural areas due to the above reasons. In rural India, at place the large size cold storage facilities are available and being used for potato storage. These cold storages are run by electricity or fossil fuels. However, in the absence of small cold storage facilities in the catchment area, farmers are forced to sell the horticultural produce in the local mandies at prevailing rate. Many times the prevailing rate is quite low in local mandies. Hence, farmers incur the loss. The horticultural produce such as mango, tomato, papaya, pomegranate, etc. are transported to long distances in hot climate without pre-cooling and the cold chain arrangement. It results quality deterioration of the product and heavy spoilage.

Solar PV Powered Refrigeration System

A cold storage facility for storage of fresh horticultural produce (6–7 tonne), powered by solar photovoltaic with battery backup has been developed at CIAE, Bhopal (Fig. 3). It consisted of PV power plant (25 kW$_p$ capacity) with battery

Fig. 3 A view of solar PV panel (*left*) and mango loading in the cold storage chamber (*right*)

bank (240 V, 900 AH) and puff insulated cold storage chamber (5 m×4.4 m×3 m) fitted with vapour compression refrigeration system (2.5 TR capacity) and a humidifier. The power conditioning unit/inverter of the solar power plant converts the DC power produced from the solar panel into three-phase AC electricity (415–420 V) for operating the cold storage unit. Temperature and relative humidity (RH) controllers were fitted in the cold storage chamber to maintain desired room temperature (5–25 °C) and relative humidity (65–95%) for storage of horticultural produce.

The detail of the system is given below.

Solar Photovoltaic (SPV) Panel System

Capacity of the SPV panel: 25 kW_p
No. of PV modules: 250 $W_p \times 100$ nos
Type of PV module: Poly crystalline silicon solar cell module (as per IEC-61215 standard).

Power Conditioning Unit (PCU)

Capacity: 25 KVA.
Inverter: AC output voltage: 415 V, 3-PH, 50 Hz, Inverter output capacity: 25 KVA, Output wave shape: Sine wave, Efficiency (at 30 °C): 85% or above
Solar charge controller: Capacity: 25 kW, Cooling: Temperature controlled fan forced cooling.
Ambient temperature range: 0–50 °C, Ambient humidity range: up to 90% condensing.

Battery Bank

Total capacity of battery bank: 900 AH at 240 V
Type of battery: Tubular lead acid solar battery,
No. of battery: 100 nos (180 AH, 12 V each), Make: Luminous.

Sensor for Temperature and RH Controller in Cold Storage

RTD sensor based temperature controller,
Temperature Range: 0–50 °C
Humidity sensor range: 0–95%.

Materials and Methods

The solar-powered cold storage system was evaluated by storage of the horticultural crop such as unripe fresh mango. The recommended temperature for mango storage is 10–13 °C (13 °C for mature green and 10 °C for partially ripe mango) and relative humidity 90–95% for maintaining post-harvest qualities by UCDAVIS Post-Harvest Technology, University of California (2014). The shelf life of the fresh matured unripe mango (Dashari varieties) was studied by storing the mango in the cold storage chamber at 12 ± 1 °C temperature and 90 ± 2% relative humidity and compared with the mango stored at ambient condition in the June month. The quality parameters such as total soluble solids (TSS), titratable acidity, firmness, physiological loss in weight of the mangoes were measured regularly during storage. Titratable acidity (TA) was determined by direct titration with 0.1 N NaOH solutions and was expressed as per cent Citric acid. The TSS values were measured by a hand refractometer and expressed as °Brix. The firmness of mango was measured with help of texture analyzer machine. The physiological loss in weight (PLW) was measured by weighing the samples regularly during the storage. The solar radiation was measured with help of the pyranometer. The energy required for operating the cold storage and energy generated from the solar power plant was measured in different months with help of the energy meters.

Results and Discussion

The average initial temperature of mango during loading was 34 °C. Temperature of the mango was slowly reduced and stabilized at about 12 °C in about 13 h. Based on the different physico-chemical parameters, it was found that Dashari

Table 1 Physico-chemical parameters of unripe mature mango during cold storage (stored at 12 ± 1 °C, and RH: 90 ± 2%)

Parameters		Dashari in cold storage	Dashari in ambient storage (control)
Safe storage life, days		15	4
Increase in total soluble solid (TSS), °Brix	At initial	8.8	8.8
	After storage	13.0	16.2
Decrease in firmness, kgf	At initial	31.6	31.6
	After storage	9.7	4.4
Decrease in titratable acidity, % citric acid	At initial	2.25	2.25
	After storage	1.0	0.92
Physiological loss in weight (wb),%		3.1	14.5%

mango stored in the cold storage (at 12 ± 1 °C temperature and RH 90 ± 2%) could be safely stored up to 15 days as compared to 4 days at ambient storage (Table 1). The ambient temperature and humidity varied between 27 and 40 °C and 40–55% respectively during the study period. The total soluble solid (TSS) of Dashari mango increased from initial values of 8.8–13.0 during the cold storage period. The TSS values indicate the development of soluble sugars in the fruits as the ripening progresses. As the fruits ripen, more carbohydrates/starches are converted into simple sugars and the TSS values increases. The loss in weight of Dashari mangoes was 3.1% in cold storage as compared to 14.5% in case of storage at ambient condition. As per some literatures, a physiological loss in weight (PLW) value of more than 10% resulted in severe shrinkage in the fruit skin that decreases its acceptance level and hence indicates the end of shelf life of the products. The cold storage was effective in maintaining the fruit firmness for a much longer period than the control ambient storage. Firmness of the mangoes was decreased from 31.6 to 9.7 kgf in the cold storage. The control samples started to ripen and softening after 2 days of storage and their firmness values decreased to 4.4 kgf 4 days of storage. The energy output from the solar PV power plant was 67–110 kWh during the sunny day in different months of the year. The average solar radiation intensity during the day (8 AM to 5 PM) varied between 366 and 712 W/m^2. The expected SPV power generation required for the cold storage unit (6–7 tonne storage capacity) was 46–68 kWh/day (depending on ambient conditions). The energy generated from the power plant was found enough to operate the cold storage system. The battery backup was used during inclement and bad weathers.

Conclusions

The solar-powered cold storage system was found suitable for storage of fresh horticultural crops for increasing shelf life of the produce. It can be a part of the cold chain arrangement to reduce post-harvest losses and improve the quality of the produce at end user. The energy output from the solar PV power plant (25 kW_p) was 67–110 kWh during the sunny day in different months of the year which was sufficient to operate the cold storage systems.

Acknowledgements The authors are highly grateful to Assistant Director General (ADG) and Director, National Science Fund (NASF/NFBSFARA), Director, CIAE and Head, AEP for valuable support and guidance for development of the Solar PV powered cold storage system and completion of the project.

References

Indian Horticultural Data base (2014). http://www.nhb.gov.in/area-pro/NHB_Database_2015.pdf
ASSOCHAM 2013. www.assocham.org
http://postharvest.ucdavis.edu/PFfruits/Mango (2014)
Energetica India (2012) www.energetica-india.net/enewsletters/enews-_2012-11-16

Feasibility of Solar Pumps for Salt Farmers

Kapil K. Samar

Abstract The utilization of solar energy to generate electric power is a prominent technology which is utilized in photovoltaic based water pumping system for agriculture. It helps to improve the agricultural productivity which improves the living standard of a farmer. India is the third largest salt producing country in the world. During the salt production, pumping of brine has to be carried out continuously, which is mostly done by diesel. A quick comparison between costs involved in running a brine pump on solar energy vis-á-vis diesel is described in this paper.

Introduction

A large number of people in developing countries still live in rural and remote area like India where the grid electricity is yet unavailable or not envisaged by the people. Pumps are critical to irrigation and communal water supply systems in rural economies. According to the U.N., agriculture accounts for 70% of global freshwater withdrawals—a harsh reality when considering the amount and consistency of power needed to obtain this water (Vick and Clark 2009). There are an estimated 21 million irrigation pumps in India out of which over 9 million runs on diesel and 12 million are on the electricity grid. Electricity consumption by irrigation pump sets alone accounts between 10 and 15% of India's total electricity consumption.

India is the third largest salt-producing country in the world, next to the United States and China. The major salt-producing states of India are Gujarat, Tamil Nadu and Rajasthan. More than 20,000 salt pan workers in Gujarat and Rajasthan currently rely on diesel pumps to earn a living (Martin 2015). Salt farming process is an extremely taxing manual process but the only mechanized phase of the salt processing is brine pumping. The pumping is done since the early days by the diesel pump resulting 70% of the total expense of salt production. This creates high

K.K. Samar (✉)
Department of Renewable Energy Engineering, Udaipur 313001, India
e-mail: Kapilsamar18@gmail.com

V.P. Singh et al. (eds.), *Energy and Environment*, Water Science and Technology Library 80, https://doi.org/10.1007/978-981-10-5798-4_13

indebtedness continuously during the farming season because farmers have to buy diesel through the credit money; hence the efficiency of the salt production going way into negative. Diesel systems are independent of natural cycles. Diesel generators cause noise, environmental pollution and are costly to operate especially as water demand rises during the growing season and fuel prices spike.

The Process of Salt Production

The salt farming process has not seen much of technological advance. It is primarily a manual process involving building of embankments, preparing the salt pans, "sowing" the salt seeds, daily "ploughing" of the pans, and the final harvesting—all done by the salt farmers themselves. The pumping of brine has to be carried out continuously during the farming season, for which farmers spend a fortune on diesel. The diesel required per engine per day is 8–10 L (Mahajan and Dangayach 2014).

Highlights of Yearly Cycle

A survey was conducted for salt farmers who live in Little Runn Kuchcha (LRK) Gujarat, India in the year 2012–13. Agariyas (salt farmers are called by this name in Gujarat) family has average 10 acres of land. Mostly farmers use 3 hp diesel pump with 10–12 h daily operation (Martin). They avoid using larger size pumps because pump needs to be replaced every after 2-year period due to corrosive nature of brine water. They start seeding for salt in the month of October. Farmers borrow money for diesel and household expenses against each "Pata" of salt assuming to almost 70–80 thousand rupees by the end of the season. In between March and April, the salt production has come to the end and each salt pan produces on an average 700 tonne of salt. Monthly expenditure and yearly income of salt farmers are described in Tables 1 and 2 respectively.

Only 8% of the Agariya's total revenues are converted into savings. Out of the total expenditure, 70% is accounted for operating the diesel pump set. Renewable

Table 1 Monthly expenses per pata (considering 3 hp diesel pumps)

Details	Amount (Rs.)[a]
Cost of diesel per month including procuring cost (9 L per day, @ Rs. 50 (Rs. 47 Diesel prices + Rs. 2.5 for transportation of fuel) per litre	13,500
Average petrol expenses for daily up down for food, water etc.	500
Total monthly expenses per pata	14,000

[a]Amount may vary from area to area

Table 2 Net revenue per pata per season (considering 3 hp diesel pumps)

Details	Amount (Rs.)[a]
Total expected revenue per pata (Rs. 230 per tonne, for 700 tonnes)	161,000
Monthly production expenses for 7 month (Rs. 14,000 × 7 months)	98,000
Monthly household expenses	1,000
Maintenance of diesel cost	10,000
Initial labour expenses	3,000
Total accrued credit per season per pata	112,000
Net expected revenue per pata	49,000
Labour and transport expenses for salt pickup	36,000
Net actual payment made against per pata	13,000

[a]Amount may vary from area to area

energy technologies promoted in the country are regarded as a means of satisfying rural energy needs of the country for operating different rural end-uses. The aim of this study is to reveal the case of solar energy for pumping with its optimal utilization, as an alternate to one of the most widely used conventional resource—Diesel.

Both DC and AC operated pumps are widely available in Indian market with almost same price. But if we study about the prices of each component, DC pump is 5–6 times costlier than AC pump. In the salt area, where there is a need to pump brine water, it is always recommended to use efficiently with low-cost pumps. AC pumps is a well-known product for the farmer that can easily be replaced by themselves at any time in comparison to DC pumps. In this investigation, a study is carried out to compare the alternatives on the basis of the economic parameters that are net present value and payback period.

Components Involved in the System

1. Solar PV array: The Solar PV array is a set of photovoltaic modules connected in series and parallel combination. The output of solar array is passed through pump controller for conditioning (Christopher et al. 2015).
2. Pump Controller: A pump controller is an electronic device that boosts the linear current. It is equipped with a maximum power point tracking (MPPT) controller and an inverter. The MPPT controls the pump as a function of solar radiation and inverter converts DC power to AC power with suitable frequency. Farmers can be benefited from the maximum amount of pump output during the day.
3. Pump Set: Pump sets generally comprise of the motor, which drives the operation and the actual pump which moves the water under pressure (Lance 2006) (Fig. 1; Table 3).

Fig. 1 Solar water pumping system for irrigation

Table 3 Technical specification of 3 hp solar pump

S. No.	Description	Parameters
1	Photovoltaic array	3900 Wp
2	Motor capacity	3 hp (AC operated)
4	Pump controller type	MPPT-based controller with inverter
5	Pump controller input and output	Input 24/48 V Output 3 phase–430 V
6	Module mounting structure	MS hot dipped galvanized
7	Water output on clear sky day	148,000 L per day from a total head of 20 metres

Operation of Solar Water Pumping System

Solar photovoltaic array directly generates DC electricity from the solar radiation. The direct current (DC electricity) is converted into alternate current through pump controller and supplied to pump for its operation. The capacity of the pump is based on total water requirement and total dynamic head. The capacity of the solar array is designed with the help of availability of yearly solar radiations on location and hydraulic energy required per day to pump the required amount of water (Dave 2016).

Solar pumping system is a practical solution to pump the brine in electric scarcity areas. Meanwhile solar pumping system can also be used in community water supply, fish farming and agriculture, forestry, and wastewater treatment engineering. The systems are also becoming more popular for use in municipal engineering, city parks, tourist sites, resorts, and even landscapes and fountains in residential areas.

Solar pumping delivers several benefits over diesel

Better return on investment
Low maintenance
No periodic tariff increases
No dependency on often unreliable grid power
No environmental pollution
Carbon credits savings

Net Present Value for Solar Versus Diesel

Net present value (NPV) at 10% discount factor for 3 hp water pump operating with diesel and solar energy is shown in Tables 4 and 5 respectively. The project analysis is done for 15 years project period. The project cost is the sum of capital cost and operating cost of diesel engine-operated pump and solar pump. The cash inflow or earning of the project comes from custom hire of irrigation service to the farmers. The hire rate is equal for both diesel pump and solar pump.

Table 4 Net present value of diesel operated pump

Year	Initial cost	Maintenance cost	Operating cost	Replacement cost	Cash flow	Yearly revenues	Present value
0	40,000				–		
1		3000	94,500		97,500	161,000	57727.27
2		3000	94,500		97,500	161,000	52479.34
3		3000	94,500		97,500	161,000	47708.49
4		3000	94,500	40,000	137,500	161,000	16050.82
5		3000	94,500		97,500	161,000	39428.50
6		3000	94,500		97,500	161,000	35844.09
7		3000	94,500	40,000	137,500	161,000	12059.22
8		3000	94,500		97,500	161,000	29623.22
9		3000	94,500		97,500	161,000	26930.2
10		3000	94,500	40,000	137,500	161,000	9060.27
11		3000	94,500		97,500	161,000	22256.36
12		3000	94,500		97,500	161,000	20233.66
13		3000	94,500	40,000	137,500	161,000	6807.11
14		3000	94,500		97,500	161,000	16721.53
15		3000	94,500		97,500	161,000	15201.4
						NPV	368130.88

Table 5 Net present value of 3 hp solar operated pump

Year	Initial cost	Maintenance cost	Operating cost	Replacement cost	Cash flow	Yearly revenues	Present value
0	224,000				–		−224,000
1		3000			3000	161,000	143636.36
2		3000		15,000	18,000	161,000	118181.82
3		3000			3000	161,000	118707.74
4		3000		15,000	18,000	161,000	97670.92
5		3000			3000	161,000	98105.57
6		3000		15,000	18,000	161,000	80719.77
7		3000			3000	161,000	81078.98
8		3000		15,000	18,000	161,000	66710.56
9		3000			3000	161,000	67007.42
10		3000		15,000	18,000	161,000	55132.69
11		3000			3000	161,000	55378.04
12		3000		15,000	18,000	161,000	45564.21
13		3000			3000	161,000	45766.97
14		3000		15,000	18,000	161,000	37656.37
15		3000			3000	161,000	37823.94
						NPV	925141.36

$$\mathrm{NPV} = -\mathrm{IC} + (\mathrm{AS} - \mathrm{AM})\left[\frac{(1+i)^n - 1}{i(1+i)^n}\right],$$

where IC is initial cost, AS is annual savings, AM is annual maintenance, i is discount rate and n is years.

NPV of solar pump is almost 2.5 times more than diesel pump which indicates that invest on solar pump is more profitable than diesel pump. Investment on solar pump is more risk-free with higher discount rate than diesel pump. Payback period of solar water pump comes to 2 years, 6 months and 11 days (shown in Table 6).

Features for Safe Future

Efficiency is one of the greatest challenges when designing a solar pumping system. The electronic controller should be such stronger that it starts the pump and keeps it running even at low sun energy input. MPPT provides uninterrupted flow, even during drastic changes in radiation. This could make cost-effective equipment for remote sites where maintenance is infrequent.

The controller's fault resets and other features should be all automatic. The controller should be automatically shut down to prevent equipment damage if the

Table 6 Payback period calculation for solar pumping system comparing to diesel pump set

Year	Total annual costs of diesel	Total annual costs of diesel	Savings due to solar	Replacement cost of solar
0	–	–	–	224,000
1	97,500	3000	94,500	–
2	97,500	18,000	79,500	–
3	97,500	3000	94,500	–
4	137,500	18,000	119,500	–
5	97,500	3000	94,500	–
6	97,500	18,000	79,500	–
7	137,500	3000	134,500	–
8	97,500	18,000	79,500	–
9	97,500	3000	94,500	–
10	137,500	18,000	119,500	–
11	97,500	3000	94,500	–
12	97,500	18,000	79,500	–
13	137,500	3000	134,500	–
14	97,500	18,000	79,500	–
15	97,500	3000	94,500	–
Payback period			2 years, 6 months and 11 days	

pump runs dry. Sensor-equipped flow measurement device gives a direct indication of performance, allowing the end-user to measure system performance on flow rather than electrical parameters.

Conclusion

Solar photovoltaic pumping systems can easily meet the irrigation requirements for land holdings for small and marginal farmers (Mike and Vicki 2002). Due to lack of grid power electricity, a large number of diesel pump sets are being deployed every year in the country (Water in a Changing World 2009). This study cleared the idea about economics of solar water pumping system against diesel pumps. It will prevent farmers from long distance travels to procure and transport diesel. From the technical perspective (reliability and easiness in operation) and economic evaluation of the technical alternatives, solar AC pumping system is found to be the most viable solution to pump brine in the salt farming areas. The payback period of solar water pumping system is 2 year and 6 months compared to diesel pump set.

Acknowledgements The author (Er. Kapil K. Samar) gratefully acknowledges Shashwat Green Fuels and Technologies Pvt. Ltd. Ahmedabad (Gujarat), India.

References

Christopher WS, Roy B, Richard W (2015) Guide to solar powered water pumping systems in New York State. New York state energy research and development authority. Accessed Sept 2015. https://www.nyserda.ny.gov

Dave D (2016) Sun power salts your food twice over. Report submitted to India climate dialogue. Accessed on 20.05.2016

Lance BBC (2006) Livestock watering handbook. British Columbia Ministry of Agriculture and Lands. Accessed Sept 2015

Mahajan S, Dangayach K (2014) Utility of solar pumps for salt farmers. Akshay Urja 8(2):31–32

Martin V (2015) Creating a social enterprise model for small salt producers in Gujarat (India) by replacing diesel pumps with solar pumps. Accessed Sept 2015. http://www.good-roots.org

Mike M, Vicki L (2002) Solar-powered livestock watering systems. Technical note submitted to ATTRA. Accessed Sept 2015. https://www.attrancat.org

Vick BD, Clark RN (2009) Determining the optimum solar water pumping system for domestic use, livestock watering or irrigation. In: Proceedings of ASES national solar conference. Buffalo, NY

Water in a Changing World (2009) The United Nations world water development report 3. Published by Unesco, London, UK

Part VI
Energy: Hydropower

Development of Uttarakhand Using Alternative Energy Source as Micro Hydropower

Jyothi Prasad and H.J. Shiva Prasad

Abstract Micro hydropower plant/water mills are local energy resource can be usefully harnessed for rural energy from the small streams or rivers. These streams usually called as gaarh are an important asset for the sustainable development of the villages of Uttarakhand. In this paper the following two micro hydropower plants of Uttarakhand is taken up for the case study and suggestions to improve the problems faced by the projects are discussed. Lacchiwala is a village situated on the foothills of Mussoorie. It is alongside NH 74 and is 20 km from Dehradun towards Haridwar. Song is the river that provides all irrigation facilities to the farmers. Apart from irrigation facility it also acts as a source of hydroelectric power. The village has a capacity of 200 villagers. Kotabagh Power Project (0.2 MW) in District Nainital in year 1989–1990 is a Micro Hydro Power Station of UPJVN Limited by the then Govt. of U.P. for exclusive development of small hydroelectric power in the state of Uttar Pradesh (now under Uttarakhand). Kotabagh Power Station is located on Dapka River, near Kalagarh—about 70 km from Haldwani and has 2 units of 100 kW each based on available discharge to the machines. Kotabagh was suffering from lack of electricity for a long time. The Kotabagh Power Station still continues to lighten Kotabagh area of the Nainital District. Development of Uttarakhand can be possible by making the villagers self dependent by developing micro power projects/water mills, so that it can be a source of power and employment opportunities for them.

Keywords Micro hydropower · Alternative energy source · Sustainable development · Employment opportunities

J. Prasad (✉) · H.J. Shiva Prasad
Department of Civil Engineering, College of Technology, G B Pant University of Agriculture & Technology, Pantnagar 263145, Uttarakhand, India
e-mail: jptce@gbpuat-tech.ac.in

V.P. Singh et al. (eds.), *Energy and Environment*, Water Science and Technology Library 80, https://doi.org/10.1007/978-981-10-5798-4_14

Introduction

Uttarakhand State is divided into Kumaon and Garhwal Divisions with 13 districts, 42 tehsils, 95 blocks and 15,689 inhabited villages and 73 towns. The State has a geographical area of 53,119 km^2 which is 1.62% of the total area of the country and supports 1.01 Crores populations which is 0.84% the total population of India (2011 census). The percentage of villages having population more than 500 is about 19% (2011 census). The hill region of districts are less developed in terms of infrastructure i.e. electricity, roads and irrigation. The inter district inequality in infrastructure leads to increasing disparity in terms of income and livelihood between the hills and the plain. Low levels of income not only result in low levels of consumption and material derivation, but also constrain human potential by restricting access to education and health facilities thereby creating a vicious cycle of poverty. In this study since Uttarakhand is rich in water resources and micro hydropower. The case study is taken up for two existing micro hydropower as an alternative energy resources of Uttarakhand to understand its performance and to promote more micro hydropower for the development of Uttarakhand.

Objective of the Study

The following are the objectives of the present study:

- Study of performance of the existing two micro hydropower projects of Uttarakhand, viz. Lacchiwala and Kotabagh Micro Hydro Project has taken up as case study.
- Remedial measures suggested to improve the problems faced by these projects are discussed.

Study Area

Lacchiwala Micro Hydro Project

Lacchiwala Micro Hydro Project situated in a village Lacchiwala has a capacity of 200 villagers, with its longitude 78° 01′ 47″ and its latitude 30° 40′ 29″ which is located on the foothills of Mussoorie, NH 74 and is 20 km from Dehradun towards Haridwar. Song is the river that provides irrigation facilities to the farmers as well as a source of hydroelectric power. This hydropower plant was setup by HESCO to provide electricity to complete area. It was established on 28 August 2005. It contains three turbines with two of them are in continuous operation. It works on simple principles and machinery. All the villagers are not taking electricity from the micro hydro project. Few villagers take electricity from power grid; hence the hydro power always runs short of its capacity. The tail race water serves as source of irrigation for the farmers and they buy electricity for their irrigation pumps and other heavy machinery. The hydropower has a capacity of 7.5 kW but runs at capacity of 5 kW only, since the energy requirement in the locality is less.

Problems Faced by the Lacchiwala Micro Hydro Project

During rainy seasons the water become violent and carries lots of sediment and boulders with it. These boulders and sediment affect the turbines hence the hydropower project has to be closed during the seasons of high discharge.

Intake

Diversion Structure

Power generation

Kotabagh Micro Hydro Power Project Description and Its Problems

Kotabagh **Micro Hydro** Power Project (0.2 MW) located in District Nainital with its longitude 79° 26′ 19″ and latitude 29° 45′ 22″. This was started during the year 1989–1990 which was established by UPJVN Limited by the Govt. of Uttar Pradesh. It is an exclusive development of small hydroelectric power in the state of Uttar Pradesh (now under Uttarakhand). This micro hydropower station having source from Dapka River, near Kalagarh—about 70 km from Haldwani, Nainital District. It is having 2 units of 100 kW each based on available discharge to the machines. Kotabagh was suffering from lack of electricity for a long time. This micro hydropower helped the Kotabagh to lighten this area. The following are the problems observed in Kotabagh project are as follows.

Sluice regulates the flow of incoming water according to demand and need. But here sluice which controls the flow to forebay tank is broken. Due to access problem to sluice, it is hard to replace or repair. The wild plants all around the sluice also interrupt the flow. There is no control on the flow, as the sluice is broken.

Trash Rack Screen checks the entry of small leaves, plants, small animals etc. which, if allowed to enter, will chock the penstock or will damage the turbine. In Kotabagh, desilting tank was at a larger distance from the forebay tank. Due to which, even after desilting, some leaves and branches can fall into the running water.

Cover Screen is provided on tank which is open, to avoid falling tree, leaves, animals and sometimes to protect people from falling in the tank. This Screen was not properly maintained on the tank. People of local village steal the steel screen, due to which animals generally falls in forebay tank and creating choking problem. No welder is ready to take the job of proper fixing of screens as there is no access and it is quite difficult to take welding equipments there.

There is no access to the structure available for inspection or repair and maintenance. In Kotabagh Project, the stairs were old and timber was used. Timber got damaged with time and no action was taken to repair the same.

BROKEN COVER SCREEN

PENSTOCK

SLUICE GATE WAS BROKEN

CHOKED SCREEN

BROKEN STAIRS

Results and Conclusion

In Lacchiwala project it has been observed that Head available at present was not sufficient, the flow of water was comparatively less, losses due to mechanical wear and tear. The channel carrying water is not lined thus results in loss of head and friction losses. Even during rainy seasons the water become violent and carries lots of sediment and boulders with it. These boulders and sediment effect the turbines hence the hydro power has to be closed for rainy seasons or during the seasons of high discharge.

Suggestions to Improve the Utilization Factor

There is a requirement of Lining of the channel for avoiding seepage, timely maintenance of turbines and other mechanical devices hence by providing sediment traps so that sediments and boulders do not enter into the turbine system.

Kotabagh Micro Hydropower plant requires overall maintenance of the project so that it can work in better manner.

Suggestions to Improve the Utilization Factor

- There is a need to construct desilting tank for settlement of silt.
- Forebay of the tank should be covered with fencing.

Development of Uttarakhand can also be possible through the alternative energy resource as micro hydropower so that rural remote hilly villages can be made self sustainable. Hence, there is a need an hour to study performance of existing Micro hydropower so that project can be made more viable.

Selected References

Annual report 2009-10 U.R.E.D.A

Final Report on "Water Mills" prepared by AHEC, IIT Roorkee for UNDP-GEF hilly hydro project

Harnessing of hydro power potential in Uttarakhand by Dr. M.C. Joshi (Additional Secretary (Energy), Govt. of Uttarakhand)

Prasad J, Shiva Prasad HJ et al (2011) Multipurpose upgraded water mill of Zameera Village of Uttarakhand: a sustainable option in hilly environment, at National Seminar, G B Pant University of Agriculture and Technology, Pantnagar, 13–15 June 2011

Prasad J, Shiva Prasad HJ, Bisht R (2011) Sustainable use of natural resources to maintain the livelihoods of the poor as Water Mill of Silmoriya Village of Nainital, Uttarakhand, at National Seminar at Apex Institute of Technology, Bilaspur Road, Rudrapur, 18 June 2011

Prasad J, Shiva Prasad HJ (2011) Resource development for hilly people—a case study of Dhokthi upgraded water mil, at National Seminar Srinagar Garhwal, 02–04 Nov 2011

Prasad J, Shiva Prasad HJ (2012) Hill area development—a case study of Bagar-micro upgraded water mill of Kumaun region, at National Seminar Srinagar Garhwal, 17–18 Mar 2012

Prasad J, Shiva Prasad HJ (2015) Performance study of KOTABAGH micro hydropower project of Uttrakhand during international conference on hydropower for sustainable development at Dehradun, India organized by Alternate Hydro Energy Centre, Indian Institute of Technology Roorkee & Govt of Uttarakhand, Dehradun during 05–07 Feb 2015

Comparative Studies on Performance of Commercially Available High-Power PC-LED Bulbs Under Tropical Conditions

Arindam Chakraborty, Amit Mukherjee, Tanmoy Das and Rajiv Ganguly

Abstract As addressed by scientists, industrialists, and policymakers from all over the globe at the **2015 United Nations Climate Change Conference**, held in Paris, France, it is a very important problem to reduce the greenhouse gas effect to limit the global temperature increase to 2 °C above pre-industrial levels. The main objectives of the conference are to reduce world pollution as well as reduce climate change in the pact, by all the nations of the world. It can be possible only by introducing eco-friendly energy solutions to achieve the goal. Light-emitting diodes play a crucial role to fulfill all the objectives. In addition to energy efficiency of LEDs, there are substantial benefits over CFL and other conventional light sources. White power LEDs show a very acceptance light level which implements some promise by saving a considerable amount of energy in indoor as well as outdoor light applications. However, the "useful life" is the most important parameter in the field of device fabrication. If the proper device is not chosen, the preferred reliability and performance will not be obtained. In this paper, two different varieties of high-power white LED with AC drivers, are put to test the "useful life" as per IESNA LN-79 Standard. In this working prototype, "useful life" of high-power LEDs as well as their thermal behavior have been studied for 6,000 h, lumen depreciation characteristics is plotted and extrapolated the data up to 50000 h. From the results obtained, the proper LED is chosen for further applications.

A. Chakraborty (✉)
Department of Electronics & Communication Engineering,
Institute of Engineering and Management, Kolkata 700091, India
e-mail: arindam.chakraborty@iemcal.com

R. Ganguly
Department of Electronics & Communication Engineering,
University of Engineering and Management, Kolkata 700156, India
e-mail: rajivganguly72@gmail.com

A. Mukherjee · T. Das
University of Engineering and Management, Jaipur 303807, India
e-mail: amukherjee31@yahoo.com

V.P. Singh et al. (eds.), *Energy and Environment*, Water Science and Technology Library 80, https://doi.org/10.1007/978-981-10-5798-4_15

Keywords White power LED · Lumen maintenance · LED reliability
Useful life · Energy efficiency

Introduction to Solid State Lighting

The increase of electricity and the development of power grids that began with Thomas Alva Edison changed the way we illuminate our lives and created a new engineering field. It is easy to appreciate the significance of light source technologies both from an economic point of view and from an environmental point of view. Light sources and lighting not only represent an economic market of billions of dollars but the utilization of energy for lighting is responsible for the production of millions of tons of carbon dioxide gas annually.

Furthermore, light are vital and light sources playing an indispensable role in daily life. The superiority of life, including aspects such as health care and urban security related to traffic and crime prevention; depend on light and its quality. The use of light sources is not limited to general lighting, but also to a range of other applications that require emissions in the ultraviolet and infrared part of the electromagnetic spectrum, such as sterilization, health science, aesthetic medicine, food processing and sterilization of hospitals or water, etc.

During the last few years, light-emitting diode technology has been used progressively more in the field of various indoor and outdoor lighting systems, like, room decoration, traffic signals, exit signs, various instrumental automation panel indicators, etc. In view of manufacturing as well as commercialization aspect, an LED system wants an accurate estimation for how long a system will keep going in the market ("life span"). However, practice of the manufacturer has to quote the life test commonly 100,000 h and it is based on an average life span of a single LED (Lumileds 2012). The useful life of an LED lighting system can be defined by the time when it fails to provide sufficient light for an intended application.

The term "solid state" refers commonly to light emitted by solid-state electroluminescence, as opposite to incandescent bulbs (which use thermal radiation) or fluorescent tubes. Compared to incandescent lighting, SSL make visible light with abridged heat generation or parasitic energy dissipation[1]. Most common "white" LEDs convert blue light from a solid-state device to an (approximate) white light spectrum using photoluminescence, the same principle used in conventional fluorescent tubes.

As we move ahead, the rest of the paper is structured as follows: Sect. 2 is based on the survey of conventional and the proposed methods. Section 3 mainly focuses on our proposed methodology, where proposed circuit diagram with principle of operation, cases considered, discusses the entire process according to which the experiment is conducted. In Sect. 4, lumen depreciation measurement of LED is

[1]www.cleanrevolution.org.

discussed. In Sect. 5, discussion on ambient and body temperature of different types of PC-LEDs is carried out. Results Obtained: trends obtained for each experiment conducted, their plot, nature and behavior are discussed in Sect. 6. In Sect. 7, concluding remarks and benefits to the industrial market and areas of improvement are discussed.

Standard Research Methodology "Useful Life" of High-Power PC-LED

In our present work, a new method of measurement of "useful life" of high-power PC-LED is proposed by measuring the forward current (for AC driving condition) of the high-power LED at a particular ambient condition and measuring the lumen depreciation over time (Bhattacharya et al. 2009) with the help of matched photo-detectors (using automated data acquisition techniques) as well as standard LUX meter (manually 8 h interval), (International Standard IESNA LM-79[2] procedure). Subsequently, the lumen depreciation characteristics are plotted. Once this is done, as in the current work, this curve can be used as a reference for predicting the "useful life" of the high-power PC-LED instantaneously just by fixing its forward current within the limits of experimental error. This can be very easily adopted by the solar PV based solid-state lighting manufacturing units of countries like India (Bhattacharya et al. 2009).

Proposed Experimental Setup and Methodology

Two numbers of white high-power LED bulbs are put to life test to determine "useful life" (Narendran et al. 2001) as per IESNA LM-79 (See Footnote 3) Standard. One from standard available manufacturers and the other is commercially available variety and matched photodiode detectors are selected. A large plywood box consisting of four chambers has been used to determine the "useful life" of the power LEDs. The size of the box is 1 m × 0.5 m × 1 m (length × width × height) (Figs. 1 and 2). The inside of the box is in deep black color to focus the light source with full brightness to the photo-detector circuit as well as the Lux meter which measures the light intensity in voltage and lux, respectively. The prime reason of the black color in the chamber is to avoid reflection of white light and also looks for resisting the light passing to outside through the box. The entire automated LED detector system is placed in a light-tight chamber to avoid the influence of ambient light. This ambient condition is chosen keeping in mind the maximum temperature condition of tropical countries like India. The entire system is backed

[2]http://apps1.eere.energy.gov/buildings/publications/pdfs/ssl/understanding_lm79_reports.pdf.

up by UPS. Each bulb has been fitted to the top of the box at the middle of each chamber. The power LEDs are constantly on burning phase with constant current driving circuit up to 6000 h to evaluate "useful life" (Narendran et al. 2001). As already mentioned for the "useful life" study, two power LED bulbs are chosen (specification, as per Table 1), LED lamps are primarily subdivided into the following two types:

(a) Company **A** standard manufacturer (3 W AC bulb with driver circuit).
(b) Company **B** commercially available varieties (3 W AC bulb with driver circuit).

The LED bulbs are being connected with constant 220 V AC with 50 Hz current driving circuits. The detail specification datasheets of the bulbs are in Table 1, in our experiment we use two types of bulbs. As per Table 1, A type, Company standard bulb is available in market and B type, is commercially available bulb in market.

In this experimental procedure the lumen depreciation of white power LEDs based on the "projecting life data", an "accelerated life test" has been accomplished to determine the "useful life" by driving them to constant current driving circuit up to 6,000 h. The "lumen depreciation data" over the burning hours has been collected, analyzed and extrapolated to 50,000 h and 20,000 h (approx.) for **A** and **B**, respectively.

As above mentioned, measurement the lux and lumen of high-power LED, we are recommended to use the standard lux meter[3] and standard own design automated current to voltage converter photo-detector circuit[4] for better accuracy. In the first case, for a portable lux meter has been used to measure the light output of the LED bulbs in lux. The digital lux meter can measure up to 50,000 lx, approximately 5,000 foot candles. It is used for checking the level of luminance; luminance is a measure of the amount of light falling on a surface. In the second case, a current to voltage converter photo detector circuit has been used for the experiment in each chamber. A op amp and a suitable photodiode are used to measure the light output as per IESNA LM 79 standard in each circuit[5].

A data acquisition system from a standard manufacturer has been interfaced with the photo-detector circuits to measure the luminous intensity. It reads analogous voltages and logs to CSV (comma separated values) files on USB pen drive (thumb drive, flash drive). The files stored on pen drive can be accessed on PC with Microsoft Excel application or any other supporting CSV format[6].

[3]http://pdf.datasheetcatalog.com/datasheet/kondenshi/EL-1KL3.pdf.

[4]http://www.electronics-tutorials.ws/io/io_4.html.

[5]http://www.sunrom.com/p/analog-data-logger-to-usb-pen-drive.

[6]http://www.aliexpress.com/store/product/Wholesale-Retail-Industrial-TM550-Digital-Non-Contact-Infrared-IR-Thermometer-With-Laser-Point-Gun-50-550/1003010_1865663721.html

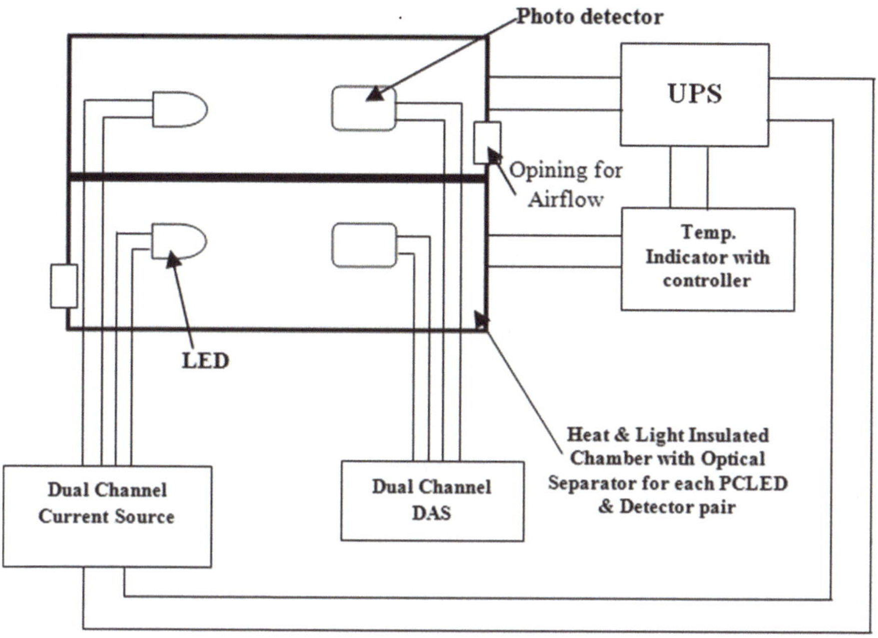

Fig. 1 Proposed schematic arrangement of High-power PC-LED's "Useful-Life" study

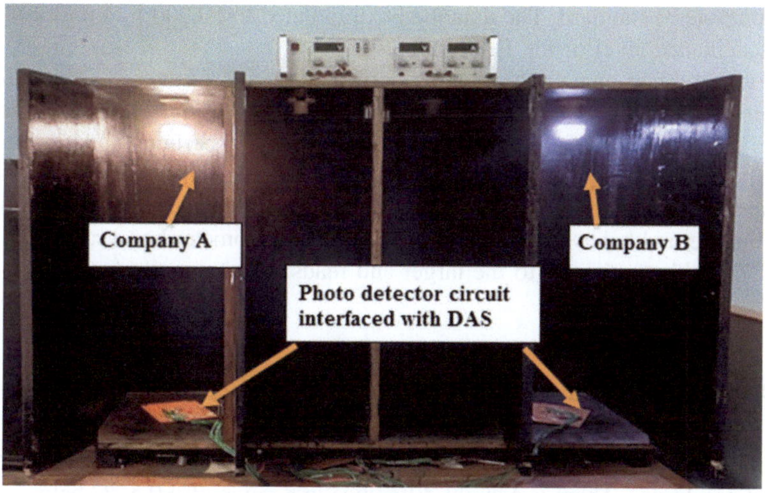

Fig. 2 Proposed experimental setup of High-power PC-LED "Useful-Life" study

Table 1 Typical high-power PC-LED specification

Specification parameters	Company **A**	Company **B**
Wattage	3 W	3 W
Lumen	270	360 (approx.)
Lumen/Watt	90	120 (approx.)
BaseType/size	B22 D	B22 D
CCT	3000 K	3000 K (approx.)
CRI	80	80 (approx.)
Average life	20000 h	Not known
Voltage	220–240 V	220–240 V

[a]http://www.aliexpress.com/item/Portable-Handheld-Luminometer-Digital-Light-Lux-Meter-Photometer-LCD-Lumen-Meter-Luminometer-Tester-Light-Measure-CG004HQ/32374958032.html

Lumen Depreciation Measurement

Now in this context, lumen depreciation data is taken between 0 h–1000 h and 1000 h–6000 h for all the high-power LED's stored in a data acquisition system, averaged and normalized considering the 100% light level as the light output after 1000 h of operation. The data for lumen depreciation is taken from 10 May 2015 to 31 December 2015 approximately 5500 h (till continue up to 19 January 2016 (≈6000 h)) for Company B. And data collection for Company A from 10 June 2015 to 31 December 2015 approximately 4500 h (till continue up to 18 February 2016 (≈6000 h)) to evaluate the "useful life" up to 6,000 h as per IESNA LM-79 (see Footnote 3) standard. The data then extrapolated to 50,000 h to understand the light acceptance level of the LED bulbs.

An infrared non-contact thermometer operated by a 9 V battery has been used to measure the body temperature of the LED bulbs as well as the ambient temperature of the high-power LED proposed "lifetime" study experimental chamber. The major portions of the LEDs such as LED panel, heat sink, driver circuit's temperature have been observed and the data has been collected in our experiment to study the thermal behavior of the bulbs. The IR thermometer has a range of −50 to +550 °C, the laser pointer to the target and reads its temperature[7].

Ambient and Body Temperature of High-Power PC-LED

The ambient temperature of all bulbs remains more than 30 °C in average (under tropical condition). There is a slight difference in a range of ±0.5 °C between them over their burning hours. The body temperature of Company A is higher than

[7]http://apps1.eere.energy.gov/buildings/publications/pdfs/ssl/lifetime_white_leds.pdf.

Table 2 Thermal characteristics of high-power PC-LED bulbs

Temperature (Avg.) (During ≈0–5,500 h)	Company A (°C)	Company B
Internal PCB of LED chip	34	35
External body of LED	45	37
Heat sink	55	NA
LED driver	60	39
Ambient temperature of chamber	31	30

Company B including its driver. There is a very little change in their body temperature over the experimented hours. So the fluctuation is negotiable and their average temperature including heat sink is given in Table 2.

Results Obtained

Figure 3 shows the lumen depreciation characteristics of High-Power PC-LEDs, the all LEDs remain their almost 90% of lumen level at the end of their burning hours. From Fig. 3 it is seen clearly that through their life test the light output of two power LEDs consider each other several times and set a linear character against time lasting their consistent and high lumen output, which is a very good parameter for lumen maintenance of power LEDs. Data for the first 1000 h are not considered, as this period is required for stabilization of the light output and this period is called the seasoning period.

Through this time span the LED bulb's light output had several ups and downs. The lumen depreciation characteristics for two bulbs is extrapolated more than 50,000 h for better understanding of their lifetime as well as to determine "useful life"(Narendran et al. 2001). It is seen from the Fig. 4 that at the end of 50,000 h the **A** remain more than 70% of their light output, whereas the **B** reaches its 60% of lumen output at only 20,000 h (approx.) which clearly establish a strong acceptable light level surpassing the L70[8] model provided by ASSIST (Chakraborty and Ganguly 2014) through their domain area in indoor as well as outdoor light applications.

Outcome and Further Research work

In this paper, a working model for Long time period to evaluate the "useful life" of white power LEDs as per IESNA LM 79 standard has been designed and interpreted in our work. The comparison of the two power LED bulbs with AC drivers,

[8]http://www.lrc.rpi.edu/programs/solidstate/assist/index.asp.

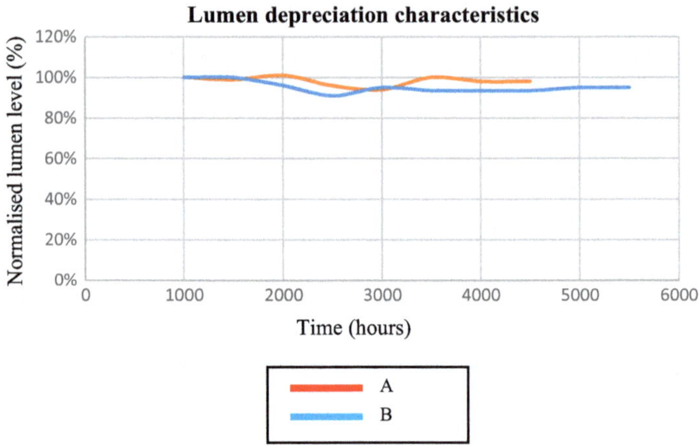

Fig. 3 Lumen depreciation plot for experimented hours

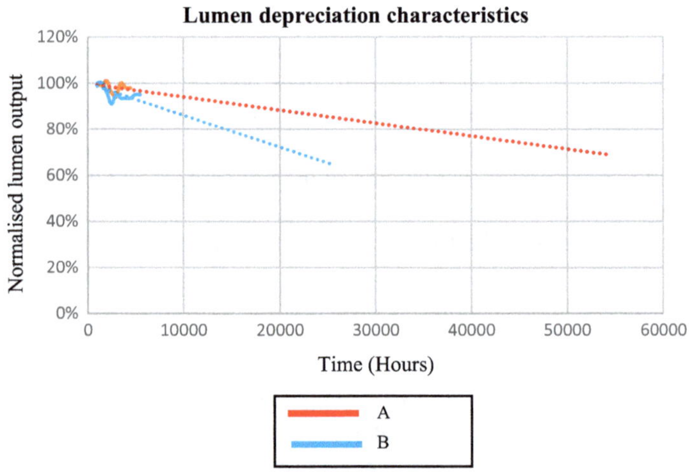

Fig. 4 Projected Lumen depreciation characteristics to estimate the "useful life' for power LEDs

have been studied and analyzed. The different lighting issues like lumen maintenance, useful life, thermal management, ambient temperature, are studied. In the present work, two types of bulb with AC drivers, one from standard manufacturer and other is commercially available put into the life tests using IESNA LM-79 standards. The results obtained from lumen depreciation characteristics are used for the selection of LED for further applications keeping in mind about the lumen maintenance, reliability, thermal management and other related issues. Subsequently, the lighting design is done dominantly for indoor application and is implemented to replace the other traditional light sources especially for CFL bulbs already existing there.

The normalized light output of **A** is more consistent than **B** and also the lumen depreciation is too less compared to other conventional light sources like incandescent and CFL as it from a standard manufacturer whereas the other is from commercially available varieties. **A** produces much heat in its body than **B** due to its color (golden yellow) and driver circuitry; **B** is bluish white and produces less heat. Overall, the light output of **A** is much consistent than **B**. From Fig. 4 it can be shown clearly that the expected "useful life" of **A** is more than 50,000 h whereas the **B** is only 22,000 h (approx.) So, it is understood from Fig. 4 that **A** is best suited for lighting applications at important places like office, showroom etc. with a good useful life (considering 70% level) while **B** may be used for other less important applications where 50% lumen depreciation level may be applicable like staircase lighting. So, in the present work, **A** from standard manufacturers is chosen for some intended applications as discussed in the subsequent sections.

References

Bhattacharya S, Chakraborty A, Ganguly R (2009) A novel technique for fast determination of 'useful life' of white phosphor-converted LED by measuring Lumen depreciation characteristics for different forward currents for solid state lighting applications. In: 18th international photovoltaic science and engineering conference and exhibition, 19–23 Jan 2009, Kolkata, India

Chakraborty A, Ganguly R (2014) Study and implementation of white power-LED based indoor lighting application for the healthcare sector. IEM OPTRONIX, Kolkata

International Energy Agency (2006) Light's labour's lost: policies for energy-efficient lighting. http://www.iea.org/publications/freepublications/publication/light2006.pdf

Lumileds P (2012) Evaluating the lifetime behaviour of LED systems. White Paper, WP15 20121218

Narendran N, Bullough JD, Maliyagoda N, Bierman A (2001) What is useful life for white light LEDs? IES Paper #52. Available: http://www.evereadyindia.com/lighting/led-filter.aspx

Design and Development of Pedal Operated Flour Mill with Multi-Applications

D. Yallappa, P.F. Mathad, Udaykumar Nidoni, T. Gururaj, R.S. Roopa Bai and Kenchappa

Abstract The pedal-operated flour mill was designed and fabricated to cater the needs of rural people, where there is shortage of electricity. In many developing countries like India there are millions of people who live a day to day life with ought reliable power to complete daily work ever increasing energy crises, and increasing the awareness to ourselves to be physically fit are the driving force for the development of pedal-operated machines for performing many day to day activities. Performance evaluation of the developed mill was done in the range of three pedaling rates, viz. 40–50, 50–60 and 60–70 rpm. It was found that 50–60 rpm of pedaling rate was optimum from quality point of view. The machine can also be used for grinding of other grains. The developed flour mill would be a boon for the farmers of remote villages and in the villages where there is scarcity of electricity or no electricity at all. The mill could be operated with a very minimum effort and even the women in the households can run the mill easily whenever required.

Keywords Grain mill · Chain drive · Grinding stones · Flour mill

Introduction

India is the largest producer of pulses in the world and also the third largest producer of cereals next to USA and China. The total food grain production in India has been reported to about 252.56 MT during the year 2011–12 under the cropping area of 126.74 million hectare. The average yield of total food grains in the year 2011–12 was 1909 kg/ha (www.Indiastat.com-2012). Among the cereals, wheat and wheat flours are the integral part of daily diet of Indian population. In India, chapatti and other variants of wheat forms the staple food of majority of the

D. Yallappa (✉) · P.F. Mathad · U. Nidoni · T. Gururaj · R.S. Roopa Bai · Kenchappa
College of Agricultural Engineering, University of Agricultural Science,
Raichur 584102, India
e-mail: yallappa.raravi@gmail.com

V.P. Singh et al. (eds.), *Energy and Environment*, Water Science
and Technology Library 80, https://doi.org/10.1007/978-981-10-5798-4_16

population. The wheat kernels are processed in flour mill (chakki) to produce wheat flour which is then used to make breads, biscuits, pasta etc. In manual process, the flour is produced by hand cranking the conventional grinding stone, but this method is characterized by strenuous, slow and low production rate. Domestic flourmills are the appliances use electrical power to grind flour from grains like wheat, corn, millet, rice, etc. Pedal power is the transfer of energy from a human source through the use of a foot pedal and crank system (Kajogbola et al. 2010). Since the thigh or quadriceps is the largest and the most powerful muscles in the human body, it makes sense to utilize it for generating energy for useful purposes

One person can generate four times more power (1/4) by pedaling than by hand cranking. Continuous pedaling at the rate of 1/4 hp can be done for only short time, about 10 min. However, pedaling at half of this power (1/8 hp) can be sustained for around 60 min. Maximum power produced with legs is generally limited by adaptations within the oxygen transportation system (David 1984). On the other hand, the capacity for arm exercise is dependent upon the amounts of muscle mass engaged and that is why a person can generate more power by pedaling than hand cranking (Tiwari et al. 2011). Pedal power enables a person to drive device at same rate as achieved by hand cranking but with less efforts and fatigue. Keeping the above facts in view, the present study was undertaken to develop a flour mill, which uses human power by means of pedal as a source of energy to drive the machine.

Materials and Methods

A concept was developed for the pedal-operated flour mill and fabrication was undertaken in the Department of Processing and Food Engineering, University of Agricultural Sciences (UAS), Raichur. The power level that a human could produce through pedaling depends on how strong the peddler is and for how long he or she needs to pedal. The power level produced by an average healthy athlete could be maximum of 75 W (Modak and Moghe 1998). A person could generate more or same amount of power for longer time, he could pedal at certain rate. A simple rule would be that most people engaged in delivering power continuously for an hour would be more efficient when pedaling rate ranging from 50 to 70 rpm (Wilson 1983). Keeping these human capabilities in mind, the proposed machine was developed, which consisted of three sub systems viz, (i) Power transmission mechanism (ii) Process unit and (iii) Out let mechanism (Figs. 1, 2, 3 and 4).

The power transmission unit basically consisted of conventional bicycle mechanism where a chain drive running over a pair of sprockets and a belt drive running over the pulley mounted on stone wheels were used. The process unit had a pair of stone wheels mounted over one another where the grains are crushed to form flour. A hemispherical bowl collector was fixed below the grinding stone to collect the flour. All these units were assembled on a supporting frame with provision to fit a seat at the top and pedaling arrangement at the bottom.

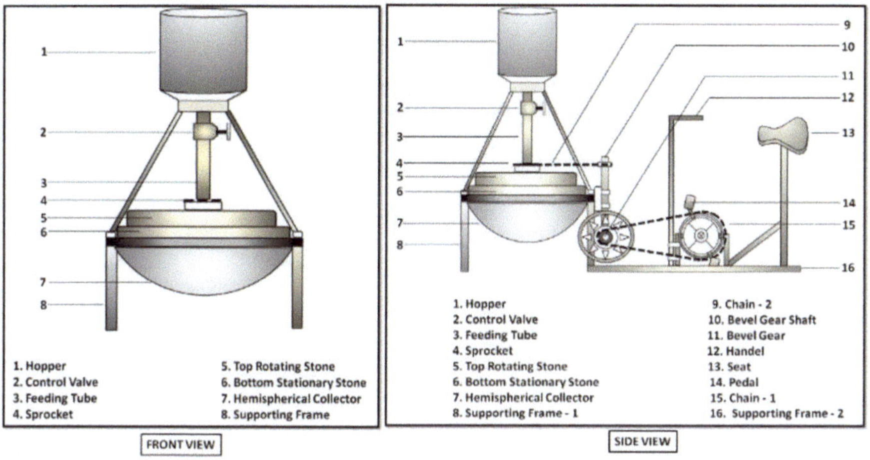

Fig. 1 Front and side elevation of pedal-operated flour mill

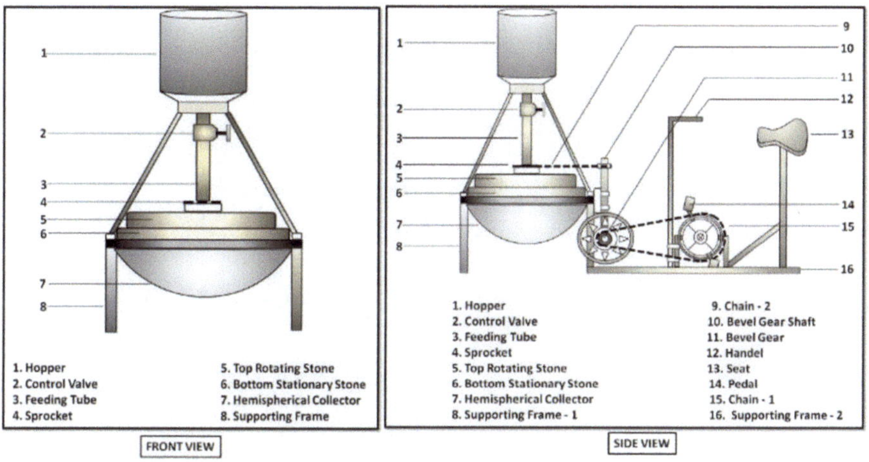

Fig. 2 3D view of pedal-operated flour mill

Hopper: A cylindrical hopper was fabricated using a mild steel sheet. The inlet and outlet diameters and slope of hopper were calculated based on the hopper capacity, where slope of the hopper played an important role in the flow of the grains.

$$V = \frac{\pi}{4} \times D^2 \times h$$

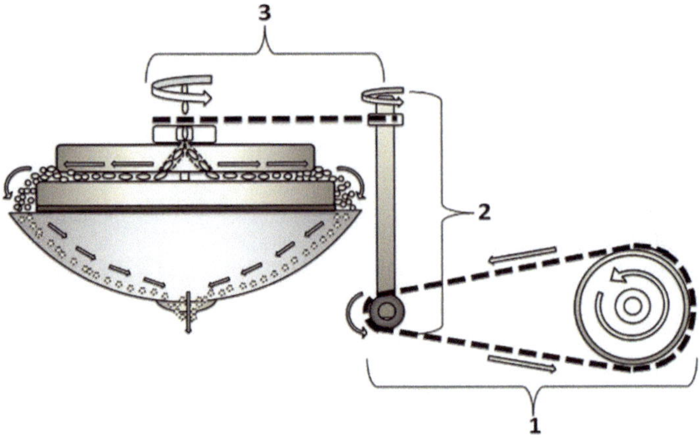

Fig. 3 Power transmission and processing unit of pedal-operated flour mill

Fig. 4 Heap of end product at the outlet of pedal-operated flour mill

where,

- *D* Top diameter of the hopper, mm
- *d* Top bottom diameter of the hopper, mm
- *h* Height of the hopper, mm,
- *s* Slant height, mm

Drive mechanism: The power was transmitted from pedal to the top stone in different stages. First the power was supplied from the pedal to big crank axle through the chain and sprocket drive. The small crank axle converted the horizontal drive to the vertical drive with the help of bevel gears. The power was transmitted from vertical drive again to horizontal drive to rotate the top stone through the chain drive mechanism. A suitable flywheel was used for smooth operation. The numbers of teeth on larger and smaller sprockets were 48 and 18 teeth, respectively, with the velocity ratio of 2.66.

Supporting frame: Mild steel angle iron of 1200 mm × 400 mm was selected for the fabrication of the frame so as to accommodate different components at the desired positions. The vertical support was provided with M. S. angle iron of 1600 mm, for the dead weight and load due to milling process. A suitable M. S. iron rods (4 No.) of 640 mm length of each were used to support the hopper on main frame. A PVC pipe (30 mm diameter) was used to deliver grains from hopper to the inlet of processing unit of flour mill. A plastic ball valve was used to control the flow rate of grains during the milling operation. M. S. flat of 3010 mm length was used to make ring for accommodating the stones on the main frame. Pedaling mechanism was fitted at the back side of the equipment at convenient position. A seat was fixed at suitable height and horizontal distance from the crank shaft for easy pedaling (Prasad et al. 2012).

Power transmission unit: The transmission of power from operator to processing unit was carried out using chain-sprocket drive. The operator used his feet and legs to pedal around the crank axel. The pedals, in turn, were fixed to a chain ring (sprocket) with teeth that engaged the bicycle's continuous chain. The chain transmitted the pedaling action to a cog on the hub of the front wheel causing the front sprocket to rotate and driven the shaft on which pulley was mounted (Kajogbola et al. 2010). This was the first stage of transmission. In second stage this power was transmitted to vertical shaft gear and stone wheels from the sprocket with the help of chain drive (Mark et al. 2006).

Processing unit: Process unit is the one where actual grinding of wheat kernels takes place. This unit consisted of stone wheel, hopper and hemispherical collector. Stone wheels made of two rigid chiseled emery stones were placed one above the other. The bottom wheel was fixed on the frame and the upper wheel was mounted on the vertical shaft to facilitate the grinding through rotational friction. A valve was provided at the bottom of the hopper to control the flow rate of grains. The selection of stone wheels were made in such a way that it should not be too heavy to cause early fatigue to the operator but it should be capable of delivering the required fineness of flour at optimum production rate (Modak and Bapat 1987). The clearance between the two stones was adjusted by changing the position of the top stone vertically. The provision was made to allow the grains inside through the hole at center of upper stone. Grains which fell on the bottom stone through the central opening were pulled along by its rotation and moved outwards by centrifugal force. At this point, grinding of grains took place in between the two stones. The ground grain (flour) was collected at the hemispherical collector provided at the bottom.

The trial experiments were conducted on a simple pedal-operated flour mill developed (Fig. 4) with 6 personnel from an age group of 25–35 years. The mean (±SD) of age, weight, height were 28.5 ± 2.5 years, 62.34 ± 21 kg, 174.8 ± 3 cm, respectively. All the trials were conducted in the laboratory where the room temperature and relative humidity varied from 29 to 33 °C and 30–40%, respectively. The machine was tested in the pedaling range of 40–50, 50–60 and 60–70 rpm (Mark et al. 2006).

Results and Discussion

At pedaling rate of 40–50 rpm, the production rate was 2.5 kg/h but the texture of flour was very fine. Average time that man can maintain the pedaling is 15 min at above pedaling rate. At 60–70 rpm, the production rate of flour was observed to be maximum but the texture of flour obtained was coarser and required regrinding for obtaining the desired texture. It was difficult to maintain a pedaling rate in the range 60–70 rpm for more than 10 min. At pedaling rate of 50–60 rpm, the production rate of flour was optimum (6.6 kg/h) and the texture of the flour was also fine. To get the required fineness double crushing is required, so overall time taken for crushing 1 kg wheat to the desired fineness is 20 min and this pedaling rate was found to be more comfortable than the other two. The similar results were reported by Hatwalne and Scholor (2012).

Conclusion

A pedal-operated flour mill was developed for the households in rural areas, where there is shortage of electricity. The developed flour mill consisted of a hopper along with a feeding tube at the top, a pair of grinding stones and the hemi spherical collector at the bottom, a seat for the operator and power transmission system. The performance of the mill was optimum at a pedaling rate of 50–60 rpm with desired fineness. The production rate with this speed was 6.6 kg/h.

References

David GW (1984) Understanding the pedal power. Ballinger Pub. Co, ISBN 0-86619-268

Hatwalne PA, Scholor ME (2012) A review on flywheel motor. Int J Eng Sci Innov Technol 1(12):190–192. http://www.indiastat.com/agriculture/2/stats.aspx

Kajogbola RA, Kadiri M, Mahamood MR, Iyanda MO (2010) Design and development of pedal powered soap mixer. J New York Sci 3(1):6–9

Modak J, Bapat A (1987) Design of experimentation for establishing generalized experimental model for a manually driven flywheel motor. In: International conference on modeling and simulation, New Delhi, India, vol 8(2), pp 127–140

Modak J, Moghe S (1998) Comparison of some bicycle drive mechanisms designed in the light of transmission angle optimization. J Papadopoulos Hypothesis Part II. In: International symposium on machines and mechanisms (ISMM '97)

Prasad A, Hatwalne M, Mahesh GM, Swapnil B, Patond M, Parvezalam I, Shaikh AN (2012) Ergonomic design of pedal operated flour mill. Int J Sci Res 4:1–3

Ricard MD, Hills-Meyer P, Miller MG, Michael TJ (2006) The effect of bicycle frame geometry on muscle activation and power during Wintage anaerobic test. J Sport Sci Med 5:25–32

Tiwari PS, Gite LP, Pandey MM, Shrivastav AK (2011) Pedal power for occupational activities effect of power output and pedalling rate on physiological responses. Int J Ind Ergon 41: 261–267

Wilson FR (1983) Bicycling science, 2nd edn. The MIT Press, Cambridge

Part VII
Energy: Biogas

Development of Family Size FRP Biogas Plant Based on Kitchen Waste

Deepak Sharma, Amol Shurpatne, Kapil Samar and Nafisa Ali

Abstract A research work has been undertaken to improve the anaerobic digestion of a family size biogas plant based on kitchen waste integrated with phase separation compartments. Laboratory experiments were carried out with developed 5 lab-scale 120 L biogas production capacity sized biphasic models. As per lab-scale study, one cubic meter size biogas plant based on FRP technology was designed developed field evaluation.

Keywords Fiber-reinforced plastic (FRP) · Nitrogen · Phosphorous Potassium (N, P, K) · Kitchen waste · Cattle dung · Biomethanation KVIC

Introduction

Kitchen waste is available from mess and canteen of industrial establishments, hotels, hostels, religious institutions etc. where food is prepared for a number of people at a time, in community type kitchen. All these produce a large quantity of kitchen waste every day, which is generally accumulated near the kitchen. This waste degrades aerobically and after sometimes they emit foul odor and make environment polluted near kitchen. The disposal of kitchen waste is a great problem for community kitchen as the waste raises a lot of problems for public health and ecology system. These wastes after some time emit bad odor, create uncongenial and attract hordes of flies which are hazardous to health (Lee and Lee 2014). Kitchen wastes contain a high amount of fat in the form of animal fat and cooking oil. This high-fat content can enhance the biogas production (Agrahari and Tiwari

D. Sharma (✉) · K. Samar
Biogas Development and Training Center, Udaipur, India
e-mail: deepshar@rediffmail.com

A. Shurpatne · N. Ali
Department of Renewable Energy Engineering, College of Technology and Engineering, MPUAT, Udaipur, Rajasthan, India

V.P. Singh et al. (eds.), *Energy and Environment*, Water Science and Technology Library 80, https://doi.org/10.1007/978-981-10-5798-4_17

2013). Disposal of kitchen waste through anaerobic digestion is most appropriate technology as it serves two purposes, first it solves disposal problem and second it produces biogas and rich manure.

Now there is worldwide awakening for protection of environment and safe disposal of food/kitchen waste. In Denmark, it is mandatory that restaurant, public institutions and other catering centers collect their food waste for recycling (Apte et al. 2013). In course of food preparation and utilization a large quantity of food wastes are produced, which may be categorized in three different forms:

1. **Uncooked food waste**: such as potato peeling, onion peeling, melon peeling, banana peeling, cabbage, leaves and stem, rotten vegetables, pumpkin peeling etc.
2. **Remain of dressed vegetables**: such as green pea, green gram etc.
3. **Cooked food**: such as rice, pieces of chapattis vegetable waste, tea leaves etc (Munda et al. 2012; Viswanath et al. 1992).

The technology has now been developed for completely prefabricated and portable HDPE, FRP, rubberized nylon fabric and RCC based biogas plants which are being made in factories and assembled on site for installation (Herrero 2011). The prefabricated biogas plants could be more suitable for rural as well as urban areas and would better meet the criteria for judging a technology ready for mass scale diffusion or commercial.

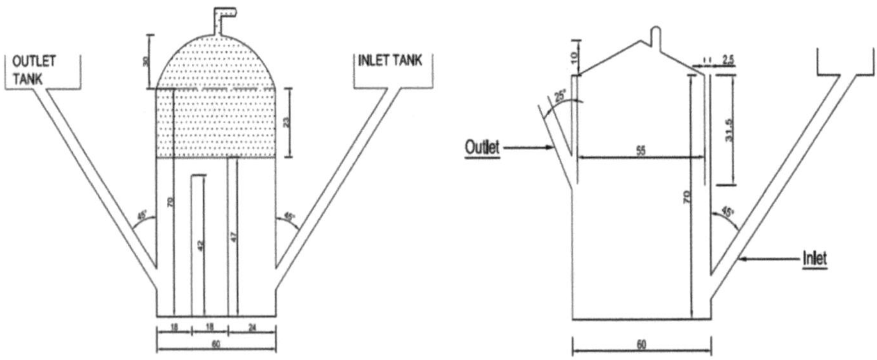

Fig. 1 Schematic diagram of lab-scale plant

Table 1 Experimental setup for laboratory study (Kaparaju et al. 2007)

S. No.	Model no.	Type	Feeding material	Quantity (%)
1	Model A	Balloon gas holder type	Cattle dung	100
2	Model B	Balloon gas holder type	Cattle dung + kitchen waste	50–50
3	Model C	Balloon gas holder type	Kitchen waste	100
4	Model D	KVIC model	Cattle dung	100
5	Model E	KVIC model	Kitchen waste	100

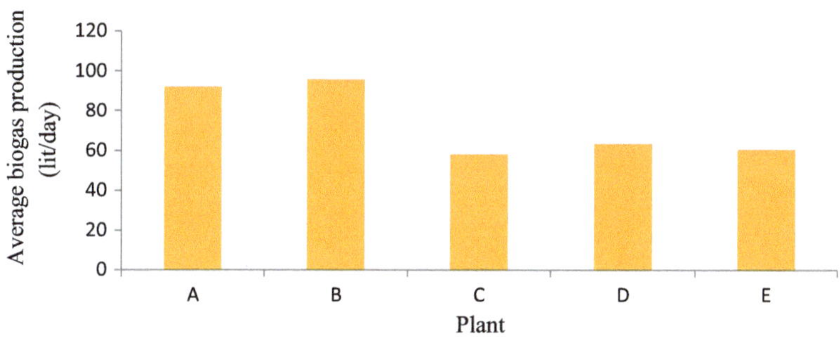

Fig. 2 Daily biogas production from lab-scale plant

Materials and Methods

The experiments were carried out in 5 lab-scale biphasic model of 250 L capacity with an internal diameter of 60 cm and height of 70 cm as shown in Fig. 1. Table 1 listing the model wise feeding pattern. The feeding was maintained once in a day (Bohn et al. 2007; Cheng et al. 2014).

Results

Daily biogas production from laboratory models was recorded. In comparison to others, plant B produces 96.08 L/day (average) during the observations 9 as shown in Fig. 2. The produced biogas comprised of 64.4% of methane and 34% of carbon di oxide. N, P and K content of digested slurry was found to be 1.42, 0.79 and 0.65%, respectively. During the study, maximum temperature was 38.7 °C and minimum temperature was 7 °C recorded (Herrero 2011).

From the laboratory study, it was concluded that the proportionate combination of kitchen waste and cattle dung in balloon gas holder type FRP plant was better. From the above results, a one cubic meter size biogas plant based on FRP technology was designed and developed at Department of Renewable Energy

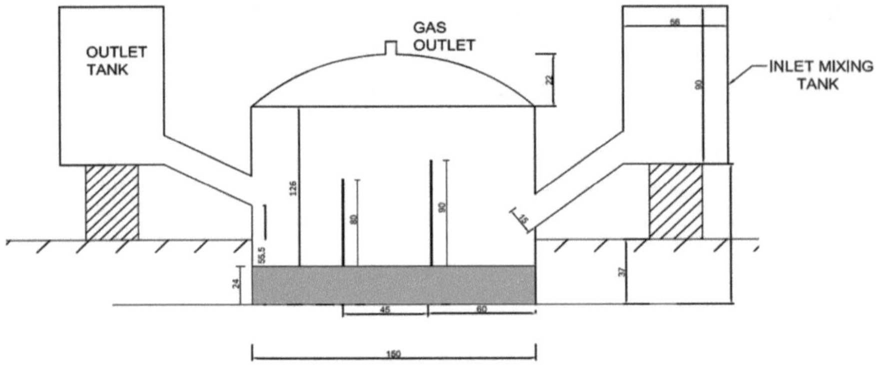

Fig. 3 Schematic diagram of family size FRP biogas plant integrated with phase separation

Table 2 Technical specification of the FRP biogas plant

S. No.	Descriptions	Unit	Specification
(A) *FRP biogas plant*			
1	Capacity of plant	m^3	1
2	Total volume of biogas plant	m^3	2.4
3	Slurry volume in digester	m^3	1.6
4	Gas storage volume with balloon	m^3	0.8
5	Diameter of digester	m	1.50
6	Depth of digester	m	1.26
(B) *Design parameters*			
1	Hydrostatic pressure	Pa	10,591.18
2	Earth pressure	Pa	756
3	Hoop stress (σ_θ)	MPa	2.21
4	Longitudinal stress (σ_L)	MPa	1.10
5	Circumferential strain		6.32×10^{-4}
6	Longitudinal strain (ε_L)		7.71×10^{-5}
7	Thickness of FRP plant	mm	6
8	Thickness of balloon	mm	0.6
(C) *Height of partition walls*			
1	First for hydrolysis stage	cm	90
2	First for acid formation stage	cm	80

Engineering, CTAE, Udaipur (Fig. 3). Technical specifications of the developed model are given in Table 2.

Biogas volume generated through developed one cubic meter biogas plant was observed daily starting from 20 April to 29 May 2015 with the help of dry biogas flow meter. Biogas production varied from 637 to 896 L/day in FRP biogas plant.

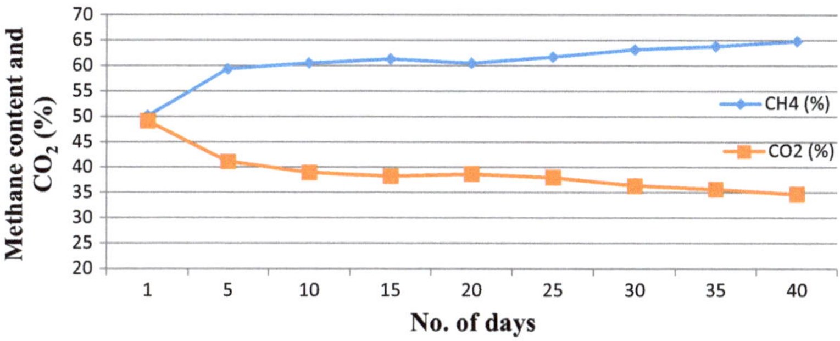

Fig. 4 Methane and carbon dioxide content of biogas in digestion period

Average biogas production from FRP plant was 786.05 L/day. The average methane and carbon dioxide content were observed as 57.43 and 39.2% respectively (Fig. 4).

Total solids, total volatile solids, pH and organic carbon content were observed at the interval of every twenty days. These parameters were observed after stabilization of methanogenic activities in the plant whereby the plant started generating near constant production of the biogas.

It can be observed that the average total solid content of the inlet charge was 9.37% in the mixture of cattle dung and kitchen waste that reduced to 7.5% for the digested slurry. Similarly, the average total volatile solid were calculated as 79.36% of total solids for the inlet charge of the plant which reduced to 64.35% after digestion (Lama et al. 2012).

The average N, P, K content in the fresh slurry was observed as 1.19, 0.77 and 0.54%, respectively, whereas the average N, P, K content in digested slurry was observed as 1.41, 0.90, 0.64%, respectively. The average N, P, K increase of 16% in the digested slurry was observed. These parameters are also suggested by Rajgopal et. al. 2013.

Conclusion

From the study it is concluded that:

- The feasibility of phase separation and duration of stages of digestion was assessed.
- Family size FRP biogas plant integrated with phase separation of 1 m^3 capacity was developed and it worked satisfactory.
- Average biogas production from FRP plant was 786.05 L/day.

References

Agrahari RP, Tiwari GN (2013) The production of biogas using kitchen waste. Int J Energy Sci 3(6):408–413

Apte A, Cheernam V, Kamat M, Kamat S, Kashikar P, Jeswani H (2013) Potential of using kitchen waste in a biogas plant. Int J Environ Sci Dev 4:370–374

Bohn I, Bjornsson L, Mattiasson B (2007) The energy balance in farm scale anaerobic digestion of crop residues at 11–37 °C. Process Biochem 42(1):57–64

Cheng C, Li Z, Mang H, Huba E, Gao R, Wang X (2014) Development and application of prefabricated biogas digesters in developing countries. Renew Sustain Energy Rev 34:387–400

Herrero J (2011) Reduced hydraulic retention times in low-cost tubular digesters: two issues. Biomass Bioenergy 35:4481–4484

Kaparaju P, Buendia I, Ellegaard L, Angelidakia I (2007) Effects of mixing on methane production during thermophilic anaerobic digestion of manure: lab-scale and pilot-scale studies. Biores Technol 99:4919–4928

Lama L, Lohani SP, Lama R, Adhikari JR (2012) Production of biogas from kitchen waste. In: Rentech symposium compendium, vol 2, pp 14–18

Lee S, Lee J (2014) Effect of sludge treatment on biogas production from *Saccharina japonica* ethanol fermentation by-products. J Ind Eng Chem

Munda US, Pholane L, Kar DD, Meikap BC (2012) Production of bioenergy from composite waste materials made of corn waste, spent tea waste, and kitchen waste co-mixed with cow dung. Int J Green Energy 9(4):361–375

Rajagopal R, Masse D, Singh G (2013) A critical review on inhibition of anaerobic digestion process by excess ammonia. Biores Technol 143:632–641

Viswanath P, Sumitradevi S, Krishana N (1992) Anaerobic digestion of fruit and vegetable processing wastes for biogas production. Bio-resour Technol 40:43–48

A Computer-Based Expert System to Design Deenbandhu Biogas Plant

Sudhir Narayan Kharpude, Deepak Sharma and Kapil Samar

Abstract Deenbandhu model is one of the most popular model used for biogas production throughout India. Because of faulty assessment of biogas plant, they have failed to deliver up to potential. The tool has been designed by developing simulation models and integrating these models with flowcharts of the design procedure. The expert system developed contains different simulation models including two main models and three sub-models. Indian standard based on family-size biogas plant designing has been used for validation of the models. The accuracy of the expert system is 98.7%. The confirmation and evaluation reported that expert system development was successful. The system developed has provided an option for economic analysis. This paper describes in detail about a computer-based expert system which can help in designing and planning of Deenbandhu biogas plant model.

Keywords Deenbandhu biogas plant · Simulation models and expert system

Highlights

- A computer-based experts system has proposed for designing Deenbandhu biogas plant.
- Simulation modelling approach has adopted for biogas plant design and material calculations.
- The expert tool has code drafted in Object-Orientated Language Visual basic.

S.N. Kharpude (✉) · D. Sharma
Department of Renewable Energy Engineering, College of Technology and Engineering, Udaipur, Rajasthan 313001, India
e-mail: sudhirkharpude@gmail.com; sudhirkharpude_ctae@hotmail.com

K. Samar
Biogas Development and Training Centre, Department of Renewable Energy Engineering, College of Technology and Engineering, Udaipur, India

V.P. Singh et al. (eds.), *Energy and Environment*, Water Science and Technology Library 80, https://doi.org/10.1007/978-981-10-5798-4_18

Introduction

The anaerobic digestion (AD) of animal wastes, biomass and wide range of other digestible organic wastes is a form of renewable energy source used worldwide. AD is a microbial action of decomposition of organic matter in absence of oxygen. Biogas has been a reasonably successful renewable energy technology developed and widely spread in India. Around 4 million family-sized biogas plants have built against the potential of 12 million plants (Chanakya et al. 2004); thus, tapping a third of potential.

Since biogas has to be one better rural energy alternative, many reasons have made an ill impact on its spread and popularity in rural part. The on-site study and grass root inspections suggest the problems have aroused from miscalculations of biogas generation volume. The faulty assessment led construction of too large biogas plants and due to lack of feed biogas plants failed to deliver (Nasery 2011).

In India, most popular biogas plant is Deenbandhu biogas plant. Deenbandhu biogas plant is a fixed dome biogas plant made using brick masonry work. The designs of Deenbandhu biogas plant are available for capacity 1–6 m^3 (Raheman 2002). In India, the volume of biogas plant model is a volume of biogas generated by biogas plant, not the digester volume.

The design of Deenbandhu biogas plant model has stemmed from for cattle dung-based biogas volume; in the case of alternative feedstocks and other animal feedstocks, we need to perform all design calculations. The biogas generation is dependent on different features like feedstocks, hydraulic retention time (HRT), feed to water ratio. So by using simulation modelling approach, it can be possible to have a number of designs available with features affecting biogas generation. Setting up digester of biogas plant needs civil construction based on various materials and the combination of different shapes and sizes. Construction of biogas digester forms a major part of investment cost (Samer 2010). So, to check financial vicinity of the beneficiary is important.

Planning and designing of biogas plant is a sophisticated work which needs multiple procedures of calculations which take long-time and many efforts; besides, making mistakes is also possible (Umaru et al. 2012). Various design procedure consists of solving polynomials which may be cubic or of fourth degree also. They take longer time and need mathematical knowledge to solve equations. Economic analysis needs deep knowledge about economic features and cost. It covers the calculation of terms like fixed cost, variable cost, capital cost, Internal Rate of Return, B:C ratio (benefit to cost ratio), payback period according to different types of materials used in biogas plant construction as well as considering biogas as a replacement for different conventional fuels.

However, by developing a tool to plan and design biogas plant, and performing all this hectic work within minutes can lessen labour costs and time needed (Samer 2010; Umaru et al. 2012). The aim of this paper was to develop an expert system for designing and planning of Deenbandhu biogas plant and to describe its utility for plant design.

Materials and Methods

General Procedure

Integrating simulation models with respective flowcharts gave the expert tool. Indian Standards for family size biogas plant, standard biogas values, and constant values of Deenbandhu model were used for back-boning of simulation models (Samer 2010). Visual Basic (version 9.0.21022.8 RTM) programming language coded the expert system program. The technique for validating, i.e. approving and evaluating models based on statistical analysis of data available in Indian Standard 9478:1989 on Deenbandhu biogas plant has been followed. Microsoft Visio flow chart creator was used to design flow of expert system.

The expert system developed contains different simulation models including two main models and three sub-models. Integrating flow charts with simulation models formed a block diagram code of the software. Main design model and Economic analysis model are the main models. Figure 1a shows the flow diagram of the computer-based expert system about main models and sub-models on one side and user on another side. The sub-models integrated with flow diagrams are having a data inflow and outflow mentored by main design model. The sub-models are as follows: (i) Deenbandhu biogas plant design sub-model which designs Deenbandhu biogas plant and its civil construction needful. (ii) Biogas generation potential assessment sub-model assess biogas generation ability and afterward finds out the volume of biogas plant for different features like feedstocks, retention time, mixing ratio, and digestion pattern. (iii) Deenbandhu biogas plant construction material sub-model which calculates the material needed for building Deenbandhu biogas plant. Sub-models go through two or more phases and so they are parted and interconnected as they want data from other sub-models. The input parameter matrix used for process flow is also as shown in Fig. 1b.

Data Collection

Development of simulation models is based on various plans, designs, features, variables and constant values of different biogas plants and their civil designs in references, mainly in (Raheman 2002; Samer 2010) and Indian standard IS 9478:1989 considered during research. The advice sought from experts from Biogas Development and Training Centre, Udaipur about designing of biogas plants considered for model interpretation and type selection of a plant to be built. The data about biogas generation potential of a different substrate according to single digestion and co-digestion collected through available research carried out mainly from Refs. (Crop Gen Database; El-Shinnawi et al. 1989; Nagamani and

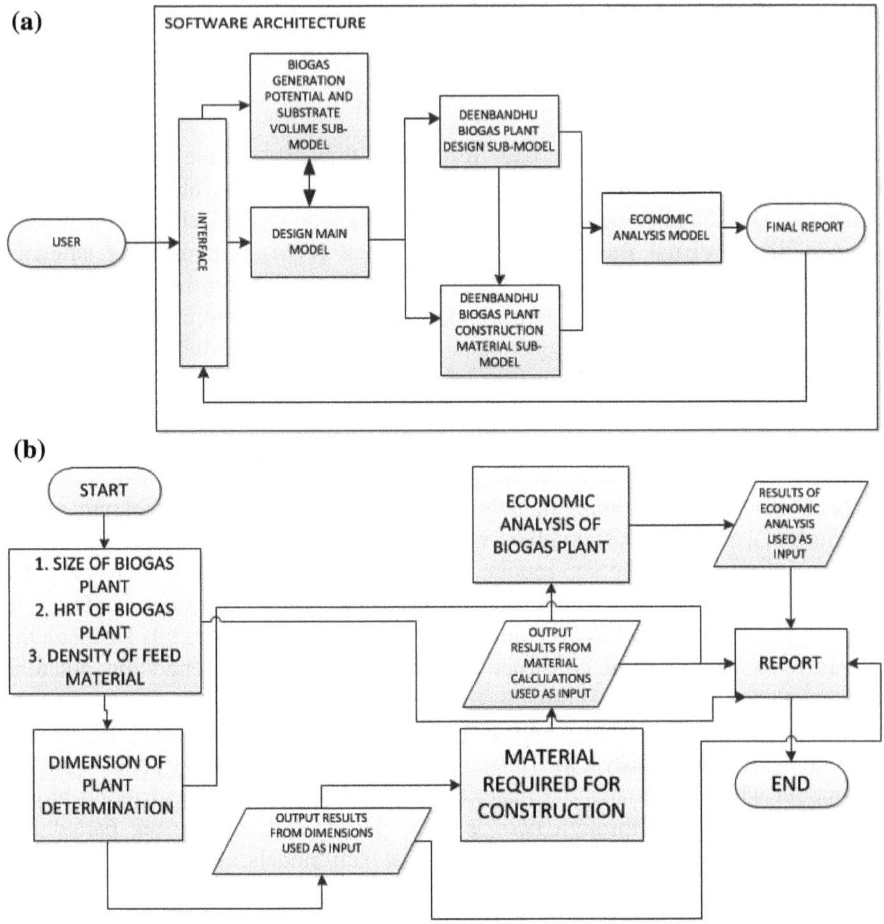

Fig. 1 **a** Flow diagram of the computer-based expert. **b** Input parameter matrix for evaluation and validation purpose

Ramasamy 1999; Annimari 2006). The cost data for economic analysis of biogas plant was collected from values used by BDTC Udaipur (Sharma et al. 2011) for calculating capital cost of Deenbandhu biogas plant.

Assumptions for Model

From the data collected and standards available according to different simulation models, some of the data and designing features were assumed while other were calculated. The assumptions (Raheman 2002) considered while designing Deenbandhu biogas model were as follows:

1. The gas holding capacity of biogas plant should be 33% of its rated capacity for plant size of 2 m^3 and above.
2. The effluent discharge opening of outlet chamber should be 150 mm lower than the bottom of the gas outlet pipe.
3. The ratio of base diameter to rise arch distance should be 5:1 for a bottom segment of the plant.
4. The plant will fail only if the stress due to the weight of mixture exceeds the strength of concrete at the bottom segment.
5. The security allowance provided for biogas plant should be 10–30% of the volume of slurry and volume of gas storage.
6. The volume of outlet should be equal to the volume of active gas storage.
7. The feed to water ratio should be 1:1 for wet digestion based on cow dung.

Validation and Evaluation of Expert System

The expert system validation aims to decide whether the system is performing correctly or not. However, evaluating an expert system aims to decide its accuracy. Standard Deenbandhu biogas plant design was tested for validation and evaluation of the expert system. The data from IS 9478:1989 was used for validation of model and software evaluation. The major dimensions of Deenbandhu biogas plant considered during this study are as shown in Fig. 2. Validation discovers if the problem is solved correctly or not. Therefore, differences determined between values picked up from the biogas plant designs in Indian standards and values calculated are used for its validation. Evaluation measures the accuracy. The percentage of calculated value to the actual value picked up from the standards, determined for each data and its average gives accuracy.

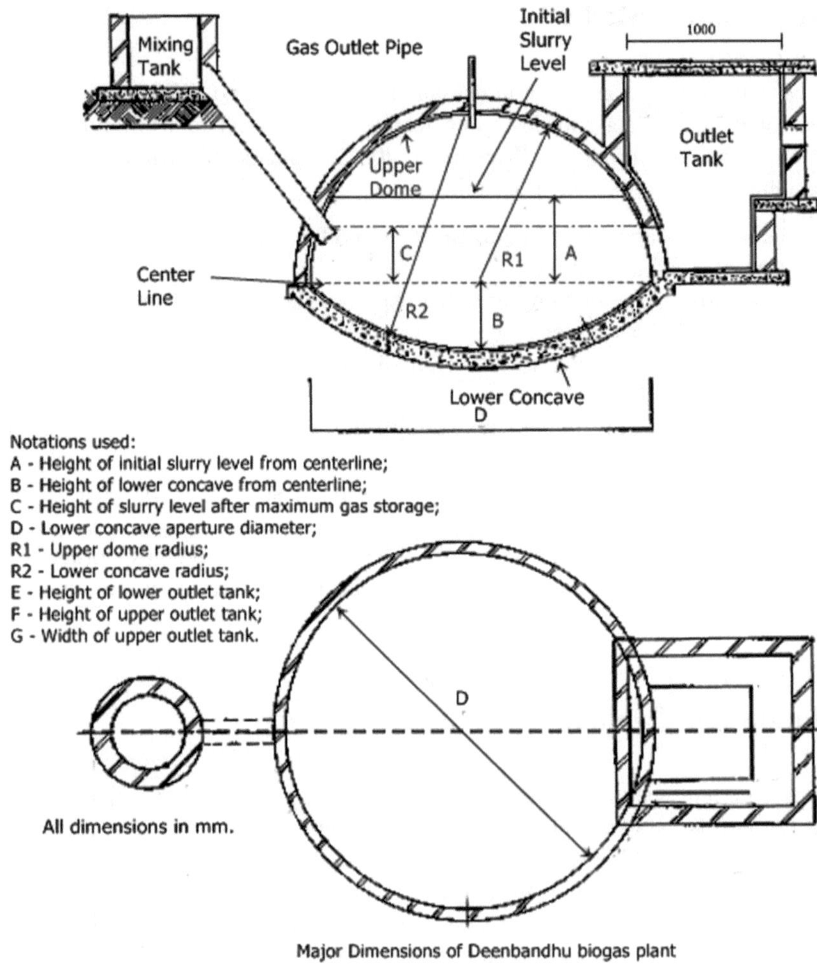

Fig. 2 Major dimensions of Deenbandhu biogas plant

Results and Discussion

The developed software could plan and design Deenbandhu biogas plant and estimate material needed to build biogas plant. It also evaluates the economic analysis of biogas plants according to various cost functions. Tables 1, 2 and 3 represent results got from the simulation. Figure 3 shows the main window for software program through which the user can enter in system. The Deenbandhu plant design interface for calculation of dimensions is as shown in Fig. 4.

Table 1 Determination of absolute variation based on actual and calculated values of Deenbandhu biogas plant dimensions for HRT 40 days (Refer Fig. 2)

Biogas plant capacity (m^3)	Parameters		*A*	*B*	*C*	*D*	*R1*	*R2*	*E*	*F*	*G*
	1	Actual	0.55	0.42	0.37	2.1	1.05	1.695	0.475	0.35	0.94
		Calculated	0.55	0.42	0.372	2.1	1.05	1.723	0.477	0.348	0.948
		Abs. variation	0	0	0.537634	0	0	1.625073	0.419287	0.574713	0.843882
	2	Actual	0.72	0.51	0.545	2.55	1.275	2.015	0.645	0.4	1.65
		Calculated	0.72	0.51	0.547	2.55	1.275	2.064	0.646	0.405	1.63
		Abs. variation	0	0	0.365631	0	0	2.374031	0.154799	1.234568	1.226994
	3	Actual	0.85	0.58	0.646	2.9	1.45	2.28	0.775	0.43	2.325
		Calculated	0.85	0.58	0.645	2.9	1.45	2.313	0.777	0.447	2.215
		Abs. variation	0	0	0.155039	0	0	1.426719	0.2574	3.803132	4.96614
	4	Actual	0.96	0.636	0.72	3.18	1.59	2.42	0.885	0.46	2.89
		Calculated	0.96	0.636	0.72	3.18	1.59	2.523	0.884	0.481	2.744
		Abs. variation	0	0	0	0	0	4.082442	0.113122	4.365904	5.3207
	6	Actual values	1.16	0.72	0.87	3.6	1.8	2.87	1.06	0.49	3.62
		Calculated	1.16	0.72	0.872	3.6	1.8	2.84	1.087	0.488	3.381
		Abs. variation	0	0	0.229358	0	0	1.056338	2.483901	0.409836	7.068915

Table 2 Determination of absolute variation based on actual and calculated values of Deenbandhu biogas plant dimensions for HRT 55 days

Biogas plant capacity (m^3)	Parameters		*A*	*B*	*C*	*D*	*R*1	*R*2	*E*	*F*	*G*
	1	Actual	0.595	0.45	0.395	2.25	1.125	1.795	0.52	0.35	0.94
		Calculated	0.6	0.45	0.387	2.25	1.125	1.851	0.521	0.379	0.871
		Abs. variation	0.833333	0	2.067183	0	0	3.025392	0.191939	7.651715	7.921929
	2	Actual	0.79	0.56	0.64	2.8	1.4	2.21	0.715	0.45	1.465
		Calculated	0.79	0.56	0.645	2.8	1.4	2.254	0.717	0.458	1.441
		Abs. variation	0	0	0.775194	0	0	1.952085	0.27894	1.746725	1.66551
	3	Actual	0.94	0.636	0.76	3.18	1.59	2.49	0.865	0.48	2.08
		Calculated	0.94	0.636	0.766	3.18	1.59	2.518	0.868	0.497	1.992
		Abs. variation	0	0	0.78329	0	0	1.111994	0.345622	3.420523	4.417671
	4	Actual	1.03	0.7	0.8	3.5	1.75	2.66	0.995	0.55	2.42
		calculated	1.03	0.7	0.843	3.5	1.75	2.749	0.963	0.562	2.349
		Abs. variation	0	0	5.10083	0	0	3.237541	3.322949	2.135231	3.022563
	6	Actual values	1.275	0.79	1.025	3.95	1.975	3.17	1.275	0.55	3.03
		Calculated	1.275	0.79	1.027	3.95	1.975	3.17	1.2	0.55	3
		Abs. variation	0	0	0.194742	0	0	0	6.25	0	1

Table 3 Determination of absolute variations for Deenbandhu biogas plant brick masonry digester material required sub-model for 40 and 55 days HRT

Biogas plant capacity	Material	Brick (Nos.)	Cement bags (Nos.)	Sand (m^3)	Stone (m^3)	Length of AC Pipe (m)	Brick (Nos.)	Cement bags (Nos.)	Sand (m^3)	Stone (m^3)	Length of AC Pipe (m)
		40 days HRT					55 days HRT				
1	Actual	700	9	2	0.8	2	800	9	2	0.9	2
	Calculated	752	9	2.05	0.8	2	811	10	2	0.9	2
	Abs. variation	6.9148	0	2.43902	0	0	1.3563	10	0	0	0
2	Actual	1100	14	3	1.3	2	1200	15	4	1.4	2
	Calculated	1154	14	3	1.3	2	1268	16	4	1.4	2
	Abs. variation	4.6793	0	0	0	0	5.3627	6.25	0	0	0
3	Actual	1420	16	4	1.6	2	1500	19	5	1.8	2
	Calculated	1420	17	4	1.6	2	1561	20	5	1.8	2
	Abs. variation	0	5.882353	0	0	0	3.9077	5	0	0	0
4	Actual	1600	22	5	1.9	2.8	1900	25	6	2.2	2.8
	Calculated	1653	21	5	1.9	2.8	1822	23	6	2.2	2.8
	Abs. variation	3.2062	4.761905	0	0	0	4.2810	8.69565	0	0	0
6	Actual	2200	28	6	2.6	2.8	2500	33	7	3	2.8
	Calculated	2250	27	6	2.6	2.8	2483	30	7	3	2.8
	Abs. variation	2.222	3.703704	0	0	0	0.6846	10	0	0	0

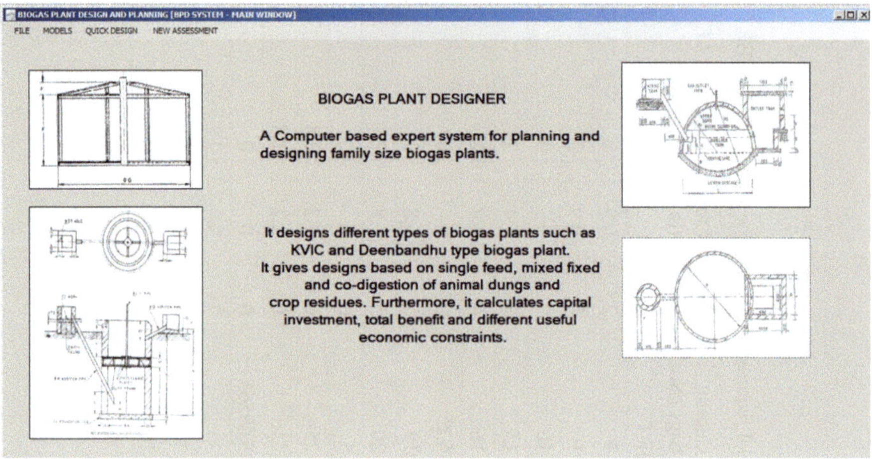

Fig. 3 Main window of expert system

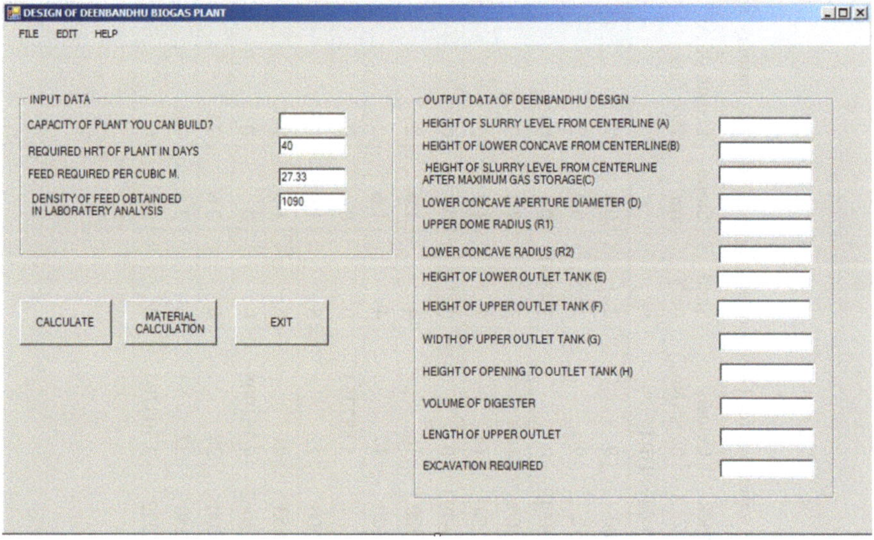

Fig. 4 Window interface for Deenbandhu biogas plant design

The report of result got through software is as shown in Fig. 5 which explains all the data needful for the design of Deenbandhu biogas plant.

Deenbandhu biogas plant design sub-model is approved and evaluated using dimensions of Deenbandhu biogas plant mentioned in IS 9478:1986 based on feed

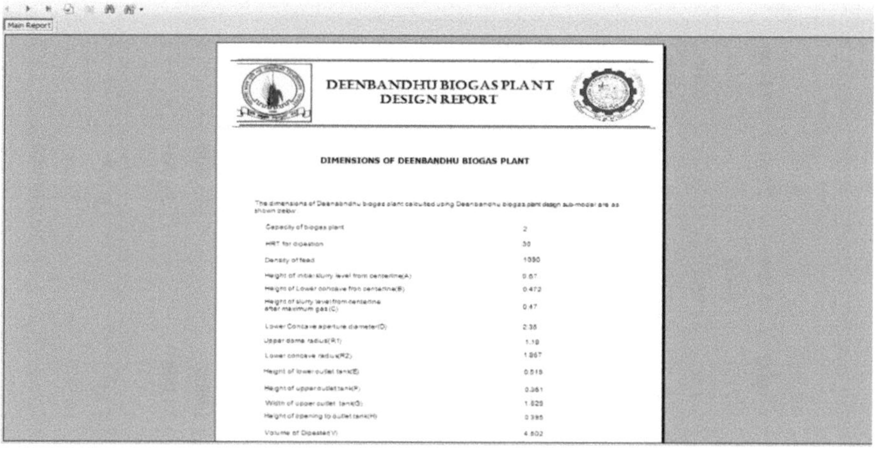

Fig. 5 Window interface for design report

to water ratio of 1:1 for wet digestion. The results for Deenbandhu biogas model of HRT 40 and 55 days are analysed and tabulated in Tables 1 and 2, respectively. The absolute variation for this sub-model varied from 0 to 7.97%. As the value of Absolute Variation was less than 10% the model is valid for use and adoption for designing Deenbandhu biogas plant. The calculated accuracy of this model was 98.8%.

Results from expert system and values collected from standards tested Deenbandhu biogas plant construction material sub-model are tabulated in Table 3. The absolute variation varied from 0 to 10%. The accuracy earned for this sub-model was 98.51%. The economic analysis of Deenbandhu biogas plant obtained from expert tool has been tabulated as shown in Table 4a–d. The detailed economic analysis results obtained revealed that tool has been successful with determining economic parameters easily and precisely. The results obtained for economic analysis have matched nearly with reference article (Singh and Sooch 2004) used for comparison.

The detailed results are as shown in Table 5. Taking all model evaluations in consideration the accuracy of the whole system found 98.7%. Absolute Variation for the whole system was not more than 10%; the models are valid for use and adoption for designing Deenbandhu biogas plant.

Summary and Conclusions

The developed expert system may plan and design Deenbandhu biogas plant according to various features used in designing. As the system tool has been successful in designing biogas plant based on cow dung; it can now be adopted for

Table 4 Economic analysis of Deenbandhu biogas plant for 40 days HRT

(a) *Fixed cost*

Material	Unit	Cost per unit	1 m^3		2 m^3		3 m^3		4 m^3		6 m^3	
Brick	Nos.	Rs. 4/Nos.	700	**2800**	1100	**4400**	1420	**5680**	1600	**6400**	2200	**11000**
Cement	Bag	Rs. 260/Bag	9	**2340**	14	**3640**	17	**4420**	21	**5460**	27	**7020**
Stone ballast	m^3	Rs. 500/m^3	0.8	**400**	1.3	**650**	1.6	**800**	1.9	**950**	2.6	**1300**
Sand	m^3	Rs. 350/m^3	2	**700**	3	**1050**	4	**1400**	5	**1750**	6	**2100**
GI pipe	Set	Rs. 200/set	1	**200**	1	**200**	1	**200**	1	**200**	1	**200**
AC pipe	m	Rs. 105/m	2	**210**	2	**210**	2.3	**240**	2.6	**275**	2.6	**275**
Iron rod	kg	Rs. 60/kg	5	**300**	7	**420**	10	**600**	12	**720**	15	**900**
Paint	kg	Rs. 250/kg	1	**250**	2	**500**	3	**750**	4	**1000**	4	**1000**
Total fixed cost				**7200**		**11070**		**14090**		**16755**		**23795**

(b) *Variable cost*

Plant size	1 m^3	2 m^3	3 m^3	4 m^3	6 m^3
Working days	10	12	16	20	30
Rate of mason	400	400	400	400	400
Cost of mason	**4000**	**4800**	**6400**	**8000**	**12000**
Rate of mason II[a]	300	300	300	300	300
Cost of mason II[a]	**0**	**0**	**4800**	**6000**	**9000**
Labours @ Rs. 250	2	2	2	2	2
Cost of labour	**5000**	**6000**	**8000**	**10000**	**15,000**
Total (in Rs.)	**9000**	**10,800**	**19,200**	**24,000**	**36,000**

(c) *Total cost*

Plant size	**1 m^3**	**2 m^3**	**3 m^3**	**4 m^3**	**6 m^3**
Fixed cost	7425.5	11,286	14,090	16,967	24,045

(continued)

Table 4 (continued)

(b) *Variable cost*					
Variable cost	9000	10,800	19,200	24,000	36,000
Total cost (in Rs.)	**16,425.5**	**22,086**	**33,290**	**40,967**	**60,045**
(d) *Economic analysis*					
Maintenance cost	**492.765**	**662.5**	**998.7**	**1229**	**1801.3**
Dung required annually (kg)	9125	18,250	27,375	36,500	45,625
Cost of dung (@ Rs. 1/kg)	9125	18,250	27,375	36,500	45,625
Annual operation and maintenance cost	**9617.7**	**18,912.5**	**28,373.7**	**37,729**	**47,426.3**
Savings by substitution					
Firewood saved annually	1277.5	2555	3832.5	5110	7665
Cost of firewood saved (Rs. 5/kg)	6387.5	12,775	19,162.5	25,550	38,325
LPG saved annually	156.95	313.9	470.85	627.8	941.7
Cost of LPG saved (Rs. 60/kg)	9417	18,834	28,251	37,668	56,502
Manure produced annually	2737.5	5475	8212.5	10,950	13,687.5
Cost of manure produced	2053.125	4106.25	6159.375	8212.5	10,265.63
Annual benefits					
On firewood substitution	8440.625	16,881.25	25,321.88	33,762.5	48,590.63
On LPG substitution	11,470.13	22,940.25	34,410.38	45,880.5	66,767.63
Annual incomes					
On firewood substitution	875.985	2074.92	3107.55	4245.99	11,429.9
On LPG substitution	3905.485	8133.92	12,196.05	16,363.99	29,606.9
Payback period (years)					
On Firewood substitution	18.75	10.64	10.71	9.64	5.25
On LPG substitution	4.20	2.71	2.72	2.50	2.02
Internal rate of return (%)					
On Firewood substitution	5.3	9.3	9.3	10.3	19
On LPG substitution	23	37	37	40	49

*Help required only for biogas plant size above 3 m^3

Table 5 Comparison of economic analysis results based on payback periods

Payback period (years)	1 m^3	2 m^3	3 m^3	4 m^3	6 m^3
Obtained results	4.20	2.71	2.72	2.50	2.02
Singh and Sooch (Singh and Sooch 2004)	4.70	2.70	2.23	1.88	1.62

designing biogas plants by varying various parameters like the feed to water mixing ratio, type of digestion pattern and different types of feed stock. Thus, by setting cow dung-based design as a bench mark, the expert tool has been designed successfully. It also evaluates design according to different feedstocks used in the digester, by calculating ability to produce biogas from those substrates. Besides, it calculates the capital investment and payback period and various economic features. Therefore, the computer-based expert system developed successfully for designing of Deenbandhu biogas plant; which offers a more useful and time-saving tool for academicians and researchers related to the field of biogas and renewable energy.

Acknowledgements This project was funded under the student project scheme by Department of Science and Technology, Government of Rajasthan. Authors wish to thank Department of Renewable Energy Engineering, College of Technology and Engineering, Udaipur for support and encouragement.

References

Annimari L (2006) Biogas production from energy crops and crop residues. Jyvaskyla Studies in Biological and Environmental Science. ISSN 1456–9701; 163. Jyvaskyla University Printing House, Jyvaskyla, 91 p. ISBN 951-39-2559-5

Chanakya HN, Rajabapaiah P, Modak JM (2004) Evolving biomass-based biogas plants: the ASTRA experience. Curr Sci 87(7):917

Crop Gen Database. http://www.cropgen.soton.ac.uk/deliverables/CROPGEN_D4_WU-soton_Methane_Database%20230511.zip

El-Shinnawi MM, El-Din MN, El-Shimi SA, Badawi MA (1989) Biogas production from crop residues and aquatic weeds. Resour Conserv Recycl 3(1):33–45

Nagamani B, Ramasamy K (1999) Biogas production technology: an Indian perspective. Curr Sci 77(1):44–55

Nasery V (2011) Biogas for rural communities. Center for Technology Alternatives for Rural Areas, Indian Institute of Technology Bombay

Raheman H (2002) A mathematical model for fixed dome type biogas plant. Energy 271:25–34. ISSN 0360-5442

Samer M (2010) A software program for planning and designing biogas plants. Trans ASABE 53 (4):1277–1285

Sharma D, Sharma A, Kurchania AK, Hakkimuddin M (2011) Training manual on biogas technology. Biogas Development and Training Center, Udaipur

Singh KJ, Sooch SS (2004) Comparative study of economics of different models of family size biogas plants for state of Punjab, India. Energy Convers Manag 45:1329–1341

Umaru S, Abubakar AF, Ismail IR (2012) Computer aided design of biogas plant for modern abattoirs. IRACST Eng Sci Technol Int J (ESTIJ) 2(3). ISSN 2250-3498

Prosopis juliflora—A Potential Problematic Weed for Lignocellulosic Ethanol Production

Vijayakumar Palled, M. Anantachar, M. Veerangouda, K.V. Prakash, C.T. Ramachandra, Nagaraj M. Naik, R.V. Beladadhi, K. Manjunatha and Beerge Ramesh

Abstract Lignocellulose-to-ethanol conversion is a promising technology to supplement starch based ethanol production. *Prosopis juliflora*, a problematic weed has been recently suggested as one of the alternative lignocellulosic biomass materials for cellulosic ethanol production. Sodium hydroxide (NaOH) pretreatment performed at 100, 120 and 140 °C in an autoclave at 15 psi, with combination of residence times (15, 30, and 60 min) and NaOH concentrations (1, 2 and 3%) indicated that almost 51% of solids were dissolved at 140 °C after 60 min pretreatment with 3% NaOH concentration. The corresponding maximum lignin reductions of 48.39, 67.01 and 74.79% were obtained at 100, 120 and 140 °C respectively for 1 h, 3.0% NaOH concentrations. Hydrolysis was carried out with CTec2® Cellulase enzyme at different loading levels (0, 15 and 30%) and the results showed that the maximum rate of saccharification (26.07 mg/g/h) was attained at 12 h for sample pretreated at 120 °C, 60 min, 2% NaOH loaded with 30% enzyme with a total maximum sugar yield of 583.9 mg/g and the carbohydrate conversion of 90.86%. Batch fermentations of enzymatic hydrolyzates carried out with 5 g/l *Saccharomyces cerevisiae* at 30 °C indicated that fermentation of 46.71 g sugar/l sample resulted in maximum ethanol of 21.84 g/l with a productivity of 0.91 g/l/h and an ethanol yield of 0.27 g/g dry biomass.

V. Palled (✉)
AICRP on Energy in Agriculture, Department of Farm Machinery and Power Engineering, University of Agricultural Sciences, CAE, Raichur 584104, Karnataka, India
e-mail: vs.palled@gmail.com

M. Anantachar · M. Veerangouda · K.V. Prakash · K. Manjunatha · B. Ramesh
Department of Farm Machinery and Power Engineering, CAE, Raichur, India

C.T. Ramachandra
Department of Processing and Food Engineering, CAE, Raichur, India

N.M. Naik
AICRP on Post Harvest Technology, Department of Processing and Food Engineering, CAE, Raichur, India

R.V. Beladadhi
Department of Soil Science and Agricultural Chemistry, Agriculture College of Engineering, Raichur, India

V.P. Singh et al. (eds.), *Energy and Environment*, Water Science and Technology Library 80, https://doi.org/10.1007/978-981-10-5798-4_19

Keywords Lignocellulosic biomass · *Prosopis juliflora* · Pretreatment Hydrolysis · Fermentation sugar yield · Ethanol yield

Introduction

One of the greatest challenges for the growing society in this century is to meet the energy demand for transportation, heating, lighting and industrial processes, which have significant impact on the environment. These challenges demanding an urgent need to carry out research work to find out the viable alternative fuels. One of the promising alternative fuels to gasoline in the transport sector is bioethanol. It is not only produced on a renewable basis from various biomass sources, including sugarcane, corn, trees, grasses, yard wastes, agricultural residues, and forestry wastes, but it also causes less net greenhouse gas (GHG) emissions during combustion. Given this reality, nations around the world are investing in alternative sources of energy, including bioethanol. Although currently most of the ethanol produced from renewable resources comes from sugarcane and starchy grains, significant efforts are being made to produce ethanol from lignocellulosic biomass (almost 50% of all biomass in the biosphere) such as agriculture residues (Bothast and Saha 1997).

For ethanol production; feedstock availability, its variability and sustainability are the main issues to be addressed. Though the generation of biomass residues in the country from agriculture is considerably large, the actual availability of a major share of these residues for bioethanol production is questionable. Thus the selection of feedstock for a future technology for lignocellulosic ethanol itself needs careful planning. With a huge population to feed and limited land availability, the nation needs to develop bioethanol technologies which use the biomass feedstock that does not have food or feed value. One of the fast growing trees which have the potential to substitute food crops for bioethanol production is *Prosopis juliflora*. It is a tree species native to Northern Mexico and the Southern U.S. that survives droughts and thrives in sunny arid regions. The plant fixes its own nitrogen, requires no seeding, fertilization or irrigation, and grows on dry, nutrient-poor soils. *P. juliflora*, a perennial deciduous thorny shrub, the common vegetation of semi-arid region of Indian subcontinent and considered to be as a problematic weed has been suggested recently to use as one of the alternative lignocellulosic biomass materials for long and sustainable production of cellulosic ethanol (Hopkins 2007). Its very nature to drought tolerance, grazing and more pellatable feed for animals like sheep and goats, could be grown in heavy sandy and saline soils of dryland tracts and not much competence to animal feed made, it could be selected as low value substrate for ethanol production. There is an urgent need for development of appropriate processing technologies to convert *P. juliflora* into a usable end product like bioethanol. Successful use of lignocellulosic biomass for bioethanol production depends on five important factors: composition of the biomass, pretreatment methods, efficient microorganisms, process integration and optimization of processing conditions.

Gupta et al. (2009) analyzed the proximate composition of *P. juliflora* wood and reported that it contained 66.20% total carbohydrate, 29.10% lignin, 2.68% moisture and 2.02% ash content. Pretreatment is a crucial process step for the biochemical conversion of lignocellulosic biomass into bioethanol. It is required to alter the structure of cellulosic biomass to make cellulose more accessible to the enzymes that convert the carbohydrate polymers into fermentable sugars (Mosier et al. 2005). Alkali breaks the intercellular bonds cross-linking hemicellulose and other compounds (lignin and cellulose), resulting in increased porosity and internal surface of biomass, and decreased crystallinity and polymerization degree of carbohydrates (Sun and Cheng 2002). Xu et al. (2010) investigated sodium hydroxide pretreatment of switchgrass for ethanol production. At the best pretreatment conditions (50 °C, 12 h and 1.0% NaOH), the yield of total reducing sugars was 453.4 mg/g raw biomass which was 3.78 times that of untreated biomass. The maximum lignin reductions were 85.8% at 121 °C, 77.8% at 50 °C and 62.9% at 21 °C all of which obtained at the combinations of the longest residence times and the highest NaOH concentrations. Sharma et al. (2013) studied the potential of potassium hydroxide pretreatment of Switchgrass for fermentable sugar production. Performer Switchgrass was pretreated at KOH concentrations of 0.5–2% for varying treatment times of 6–48, 6–24, and 0.25–1 h at 21, 50, and 121 °C, respectively. They reported that the pretreatments resulted in the highest percent sugar retention of 99.26% at 0.5%, 21 °C, 12 h while delignification up to 55.4% was observed with 2% KOH, 121 °C, 1 h. The fermentation of both acid and enzymatic hydrolysates, containing 18.24 and 37.47 g/L sugars, with *Pichia stipitis* and *Saccharomyces cerevisiae* produced 7.13 and 18.52 g/L of ethanol with corresponding yield of 0.39 and 0.49 g/g, respectively (Gupta et al. 2009). Keeping the above points in view, the present paper investigated the potential of *P. juliflora*, a problematic weed for lignocellulosic ethanol production.

Materials and Methods

The stems of *P. juliflora* wood available in the University of Agricultural Sciences, Raichur campus, Karnataka, India, were harvested up to 6 in stubble and were dried in open sun for about a week. The wood were cut into small chips and oven dried at 70 °C in a forced air oven in cloth bags for 72 h. Then oven dried samples were ground to pass through a 2-mm sieve in a hammer mill (Crompton Greave Ltd., NDA 2 TOP) and stored at room temperature in zip-locked plastic bags for use in further studies.

The initial composition of *P. juliflora* was analyzed using Laboratory Analytical Procedures (LAP) adopted by National Renewable Energy Laboratory (NREL) for the measurement of total solids, acid-insoluble lignin (AIL), acid-soluble lignin (ASL) and ash content (Sluiter et al. 2005a, b; Sluiter et al. 2008). While the structural carbohydrates (cellulose and hemicelluloses) represented by total reducing sugars of biomass was estimated by the 3,5-dinitrosalycylic acid (DNS) method (Ghose 1987; Miller 1959). In this experiment, sodium hydroxide (NaOH) pretreatment of *P. juliflora* woody substrate at different elevated temperatures ranging from 100 to 140 °C with various combinations of residence times and NaOH concentrations was explored. Pretreatment of *P. juliflora* woody substrate samples were performed at 100, 120 and 140 °C in an autoclave at 15 psi, with residence times of 15, 30, and 60 min each. All the temperature–time pretreatment combinations were performed with sodium hydroxide (NaOH) concentrations of 1, 2 and 3% (w/v) in a 3^3 factorial complete randomized block design.

Five grams of *P. juliflora* woody substrate samples was mixed with 50 ml of NaOH solution in 125 ml bottles using glass rods, and the bottles were sealed and kept in an autoclave. The pretreated samples were filtered through pre-weighed filter paper (*Whatman* filter paper No. 1) in vacuum flask using a vacuum pump. The bottles were rinsed with 50 ml DI water to recover the residual solids. All solids accumulated on the filter papers in the filtration set up were quantified by oven drying and considered in solid recovery calculations. Approximately 5 g of wet biomass was drawn from each pretreated sample and dried at 105 °C in conventional hot air oven for estimation of solid recovery. A similar amount was placed for vacuum drying at 40 °C in vacuum oven to obtain samples for estimation of acid-insoluble lignin to study the effect of pretreatment conditions on delignification of *P. juliflora*. Filtrate from the AIL acid hydrolysis was utilized to study the effect of sodium hydroxide (NaOH) pretreatment on reducing sugar content generated in each pretreated sample at various temperature–time combinations (100–140 °C and 15–60 min) using 1–3% NaOH concentrations.

Hydrolysis was carried out at 8% solid loading (of total volume 20 ml) to examine the effect of enzyme loading levels (0, 15 and 30%) on the untreated sample and selected pretreated samples for fermentable sugar production with a 3 × 4 factorial design. The CTec2® Cellulase enzyme complex sponsored by Novozymes, Beijing, China was used for conducting research on hydrolysis of samples for fermentable sugar production. The enzyme complex was reported to have an activity of 108.3–168.8 floating-point unit/ml (Eckard et al. 2012; Kodaganti 2011) and protein content 117–185.2 mg protein/ml (Eckard et al. 2012; Eylen et al. 2011).

To generate enough biomass for hydrolysis at the various conditions, pretreatments were performed in six replicates and two replicates each were combined randomly and mixed thoroughly to generate three larger replicates. This was done to avoid the impact of any scale changes during pretreatment of larger amounts. Untreated samples with equivalent enzyme loading were also hydrolyzed as control. Pretreated and untreated samples with no enzyme were prepared to determine the effect of soaking. Hydrolysis was performed for 72 h at 50 °C in a shaking water bath (KEMI Make) at 150 rpm. The Laboratory Analytical Procedure (LAP) adopted by National Renewable Energy Laboratory (NREL) for enzymatic saccharification of lignocellulosic biomass (Selig et al. 2008) was followed for conducting enzymatic hydrolysis.

The samples were withdrawn at regular intervals of 12 h and centrifuged at 4,000 rpm for 10 min in a high speed refrigerated centrifuge (KEMI Make), and the filtrate was collected for sugar analysis. Upon termination of hydrolysis, the samples tubes were kept in high speed refrigerated centrifuge for centrifugation. The filtrate was collected and stored separately at 4 °C in a freezer for estimation of total sugar generated. The fermentable sugars generated during the hydrolysis were estimated by 3,5 dinitrosalycylic (DNS) acid method (Ghose 1987; Miller 1959).

The saccharification rate at regular intervals was calculated using the formula;

$$\text{Saccharification rate}\,(\text{mg/g/h}) = \frac{\text{Sugar yield}\,(\text{mg/g dry biomass})}{\text{Saccharification time}\,(\text{h})} \quad (1)$$

Further, carbohydrate conversion was calculated using the following formula (Gupta et al. 2009);

$$\text{Carbohydrate conversion}(\%) = \frac{\text{Reducing sugar concentration obtained}}{\text{Potential sugar concentration in the substrate}} \times 100 \quad (2)$$

Based on maximum total sugar yield obtained and highest carbohydrate conversion, the sample pretreated at optimal conditions (120 °C, 60 min, 2% NaOH) and hydrolyzed with 30% enzyme loading was selected for subsequent fermentation for bioethanol production. Batch fermentations of enzymatic hydrolyzates were carried out in 500 ml Erlenmeyer flask incubated with 5 g/l *S. cerevisiae* at 30 °C (Thuesombat et al. 2007). The yeast culture was obtained from the Department of Processing and Food Engineering, College of Agricultural Engineering, UAS, Raichur. The samples were withdrawn at regular intervals of 6 h and centrifuged at 10,000 rpm for 15 min at 4 °C in a high speed refrigerated centrifuge, and the filtrate was collected and saved for subsequent ethanol estimation. Upon termination of fermentation for 36 h, the samples tubes were centrifuged at 10,000 rpm for 15 min and the filtrate was collected for estimation of total ethanol yield. Ethanol concentration present in the fermented sample was estimated by titration method. Simultaneously, the samples were withdrawn at regular intervals of 6 h and analyzed for sugars (Gupta et al. 2009). All treatments in this study were conducted in triplicate. Design expert-7 Software was used for data analysis at 99% confidence level.

Results and Discussion

The results pertaining to the composition of biomass, effect of pretreatment, enzymatic hydrolysis of selected samples and fermentation of hydrolyzate are presented and discussed in the following sections.

Composition of Biomass

The initial composition of *P. juliflora* was analyzed and the results are presented in Table 1. The total solids content present in *P. juliflora* woody substrate selected for the study was 98.80%. This is in agreement with the moisture content (2.68%) reported by Gupta et al. (2009) which represents total solids. The average acid-insoluble lignin (AIL) of *P. juliflora* was observed to be 30.18% which is in close agreement with the lignin present in *P. juliflora* (31%) reported by Rajput and Tewari (1986). Similar results (29.10%) in *P. juliflora* were reported by Gupta et al. (2009) and 20–32% in dry wood (Alriksson 2006). The total carbohydrate portion (cellulose and hemicelluloses) represented by total reducing sugar content of *P. juliflora* was reported to be 64.26%. This is in close agreement with the findings of Gupta et al. (2009) who reported that *P. juliflora* wood contained total carbohydrate of 66.20%. Also, the total carbohydrate of *P. juliflora* pod flour (69.20%) reported by Choge et al. (2007) was in agreement with present findings. Ash content present in *P. juliflora* sample was 2.01% which is fairly close in agreement with the ash content of *P. juliflora* (2.02%) reported by Gupta et al. (2009).

Effect of Alkaline Pretreatment

The results pertaining to the effect of sodium hydroxide (NaOH) pretreatment at various temperature–time combinations (100–140 °C and 15–60 min) using different concentrations ranging from 1 to 3% NaOH on solids recovered after each pretreatment are presented in Table 2.

Solid recoveries after each pretreatment ranged between 49.77 and 91.43%. It was observed that the solids recovered were maximum (91.43%) in the sample pretreated at 100 °C, 15 min combination with 1% NaOH, whereas, a minimum of 49.77% was observed in the sample pretreated with 3% NaOH at 140 °C, 60 min. As the severity of pretreatment increased in terms of higher temperature, treatment time and NaOH concentration; the solids recovered in the sample decreased. The main effect of both treatment time and NaOH concentration had significant ($p < 0.01$) impact on solid recovery. The interaction effect between temperature and concentration had a significant ($p < 0.01$) impact on loss of solids and also, the combined effect of all the three factors had significant effect on solids recovery.

Table 1 Composition of *Prosopis juliflora*

S. No.	Component	Dry weight (%)
1	Total solids	98.80 ± 0.71
2	Acid-insoluble lignin	30.18 ± 0.33
3	Acid-soluble lignin	1.67 ± 0.14
4	Carbohydrates (total sugars)	64.26 ± 0.53
5	Ash	2.01 ± 0.23

Table 2 Effect of sodium hydroxide (NaOH) pretreatment on solids recovery of *Prosopis juliflora*

		Solids recovery (%)		
Temperature (°C)	Time (T-min)	NaOH concentration (%)		
		1	2	3
	15	91.43 ± 0.36	87.78 ± 0.59	79.88 ± 0.62
100	30	87.41 ± 0.51	83.23 ± 0.70	75.42 ± 0.38
	60	74.34 ± 1.21	68.85 ± 0.35	64.31 ± 1.84
	15	78.50 ± 0.72	73.90 ± 0.69	66.98 ± 0.24
120	30	74.54 ± 0.69	70.59 ± 0.86	64.52 ± 0.70
	60	72.21 ± 0.49	66.65 ± 1.06	62.94 ± 0.32
	15	74.40 ± 1.18	70.08 ± 0.28	63.88 ± 0.81
140	30	70.47 ± 0.42	64.76 ± 0.45	57.59 ± 0.17
	60	63.16 ± 0.53	58.99 ± 0.85	49.77 ± 0.56

KOH pretreatment of switchgrass followed a similar trend of increased solid loss with increasing intensity of treatments in terms of temperature and high concentration (Xu et al. 2010). Also, the results are in agreement with the findings of Sharma et al. (2013) for KOH pretreatment of switchgrass.

The results pertaining to the effect of sodium hydroxide (NaOH) pretreatment on acid-insoluble lignin (AIL) reduction using different concentrations ranging from 1 to 3% NaOH at various temperature–time combinations (100–140 °C and 15–60 min) are shown in Fig. 1a–c. At 100 °C, a maximum AIL of 28.25% was observed in the sample pretreated with 1% NaOH concentration for 15 min, while it was minimum (16.30%) in the sample pretreated for 60 min with 3% NaOH concentration. As the NaOH concentration and pretreatment time increased, the acid-insoluble lignin decreased (Fig. 1a). Acid-insoluble lignin in the samples pretreated at 120 °C ranged between 10.42 and 22.10% (Fig. 1b). As the severity of pretreatment time and NaOH concentration was more, the acid-insoluble lignin present in the sample was less. A maximum of 22.10% was observed in the sample pretreated with 1% NaOH concentration for 15 min, while it was minimum (10.42%) in the sample pretreated for 60 min with 3% NaOH concentration. Pretreatment with 2% NaOH concentration for 60 min resulted in 59.72% delignification at 120 °C which is in close agreement with the results of Silverstein et al. (2007) that NaOH pretreatment resulted in the highest level of delignification (65.63% at 2% NaOH, 90 min, 121 °C). At 140 °C, AIL after pretreatment ranged between 7.96 and 18.78% (Fig. 1c). It was observed that the maximum (18.78%) AIL was observed in the sample pretreated with 1% NaOH concentration for 15 min, while a minimum of 7.96% was recorded in the sample pretreated for 60 min with 3% NaOH concentration. About 10.54–74.79% of the original acid-insoluble lignin was lost during various combinations of pretreatments depending on severity. The corresponding maximum lignin reductions of 48.39, 67.01 and 74.79% were obtained respectively at 100, 120 and 140 °C, for 1 h,

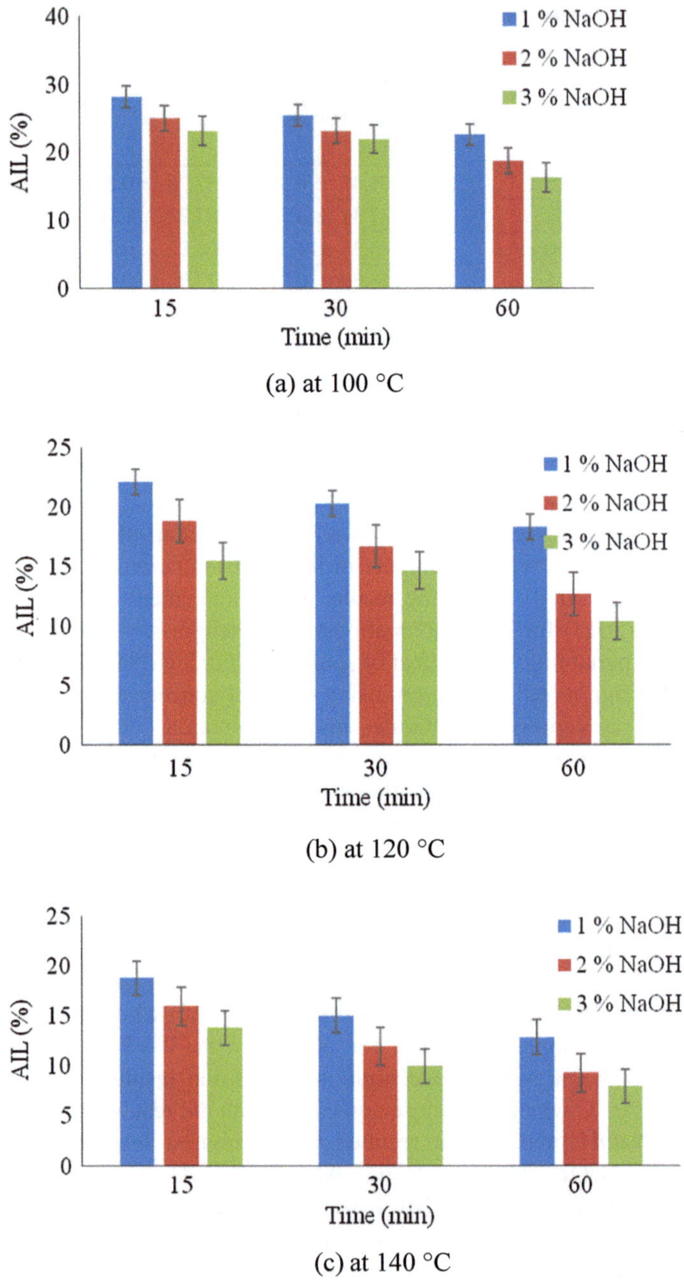

Fig. 1 Acid-insoluble lignin (AIL) of *P. juliflora* pretreated with 1.0–3.0% NaOH at different temperature–time combinations

Table 3 Effect of sodium hydroxide (NaOH) pretreatment on release of reducing sugar content of *Prosopis juliflora*

		Reducing sugar content (mg/g dry biomass)		
Temperature (°C)	Time (min)	NaOH concentration (%)		
		1	2	3
	15	339.40 ± 7.35	361.41 ± 7.46	349.99 ± 5.80
100	30	353.11 ± 1.75	392.73 ± 3.36	371.80 ± 4.48
	60	385.17 ± 7.20	394.19 ± 2.83	390.65 ± 2.94
	15	353.57 ± 8.37	378.72 ± 12.31	372.15 ± 12.62
120	30	361.93 ± 2.44	406.40 ± 4.21	392.38 ± 6.37
	60	398.60 ± 4.51	418.29 ± 3.22	389.38 ± 8.22
	15	356.33 ± 2.54	354.29 ± 4.35	348.03 ± 1.84
140	30	345.04 ± 3.82	337.03 ± 1.83	317.33 ± 1.46
	60	336.87 ± 9.14	324.77 ± 6.59	295.73 ± 3.35

3.0% NaOH concentrations. Maximum lignin reductions at different temperatures were all obtained at the combinations of highest NaOH concentrations and longest treatment times, which indicated a close relationship between pretreatment severity and lignin reduction. The similar results were reported by Xu et al. (2010), Sharma et al. (2013).

The results pertaining to the reduction in AIL were statistically analyzed and it was observed that the main effect of both treatment time and NaOH concentration had significant ($p < 0.01$) impact on lignin reduction at all the three temperatures. The interaction effect between temperature and concentration also had a significant ($p < 0.01$) impact on delignification. However, the combined effect of all the three factors on acid-insoluble lignin was not significant at 1% level of confidence. Statistical analysis indicated that at 120 and 140 °C, residence time had a significant impact ($p < 0.01$) on lignin reduction at all three NaOH concentrations. However, at 100 °C, residence time had significant impact ($p < 0.01$) on delignification of sample at higher concentrations.

On an average, reducing sugar (RS) content generated ranged between 295.73 and 418.29 mg/g dry biomass (Table 3) after pretreatment at different temperature–time combinations using various concentrations of NaOH.

It was observed that when the sample was pretreated with different concentrations of NaOH, at 100 and 120 °C, the release in sugar increased with increase in acid concentration up to 2.0% (v/v) NaOH and it declined thereafter. While at 140 °C, the acid concentration beyond 1.0% (v/v) resulted in continuous decrease in release of sugar as the pretreatment increased. The maximum sugars of 418.29 mg/g were released, when the sample was pretreated with 2.0% NaOH concentration at 120 °C for 60 min. Whereas, it was minimum (295.73%) in the sample pretreated at 140 °C, 60 min combination with 3% NaOH (Table 3). The highest carbohydrate retention of 65.09% was observed in the sample pretreated at 120 °C with 2% NaOH and 60 min. These results were more pronounced than the sugars retained (60.6%) in KOH-pretreated samples of switchgrass (Sharma et al. 2013) and lesser than the sugars retained (74.8%) in NaOH pretreated switchgrass (Xu et al. 2010).

The effect of all the factors on the reducing sugar content was statistically analyzed. The main effect of temperature, time and concentration was significant ($p < 0.01$) on reducing sugar released. Also, their interaction effect and combined effect of all the factors were significant ($p < 0.01$) on carbohydrate availability. It was observed that 34.90–53.98% of the original untreated reducing sugar content was lost during various combinations of pretreatments depending on severity.

Selection of Optimal Pretreatment Conditions

The upper and lower limits of all the factors and also, the minimum and maximum values of solids recovered, acid-insoluble and soluble lignin and reducing sugars generated were given as input to the statistical software. Selections were based on maximum solids recovery, minimum AIL, i.e., maximum delignification and maximum carbohydrate (reducing sugar) retention.

The desirability index of solids recovery, acid-insoluble lignin, acid-soluble lignin and reducing sugar content retention were found to be 0.9935, 0.7348, 0.4009 and 0.9719, respectively for pretreatment combination of 2.0% NaOH, 60 min at 120 °C which has highest combined desirability of 0.7303 (Fig. 2). Hence, this pretreatment combination of 120 °C, 60 min, 2.0% NaOH was chosen as the optimum for further enzymatic hydrolysis for fermentable sugar production.

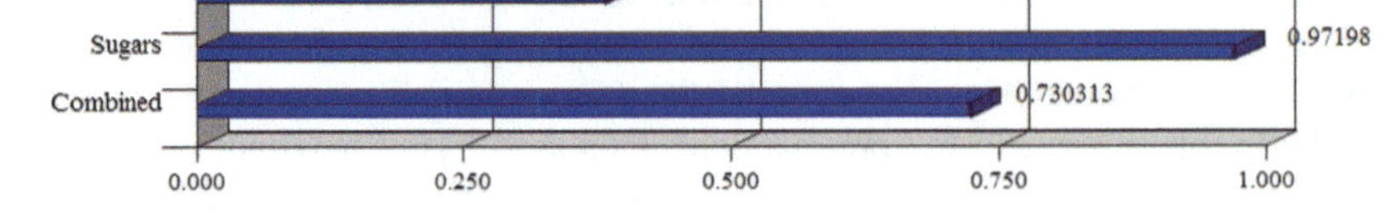

Fig. 2 Desirability index of solids recovery, lignin reduction and reducing sugar retention for optimal pretreatment condition (120 °C, 60 min, 2% NaOH)

Enzymatic Hydrolysis

The results pertaining to total sugar yield in the hydrolyzate of the untreated and selected pretreated samples with various enzyme loadings levels of 0, 15 and 30% are presented in Table 4.

The total sugar yield in the hydrolyzate of untreated and pretreated samples ranged from 26.4 to 583.9 mg/g biomass. A maximum total sugar yield of 583.9 mg/g biomass was recorded in the hydrolyzate obtained from the sample pretreated at optimal conditions (120 °C for 60 min with 2% NaOH) when loaded with 30% enzyme. Whereas, it was minimum (26.4 mg/g biomass) in the hydrolyzate of the sample pretreated at 140 °C for 30 min with 1% NaOH without enzyme loading (Table 4). The total sugar yield increased when the enzyme loading level increased for all the samples hydrolyzed. The maximum sugar yield obtained (583.9 mg/g biomass) at optimal pretreatment conditions of 120 °C for 60 min with 2% NaOH loaded with 30% enzyme in this study were in close agreement with maximum sugar yield of 582.4 mg/g biomass obtained from hydrolyzate of switchgrass pretreated at optimal conditions of 0.5% KOH, 12 h at 21 °C with same (CTec 2) enzyme loading level of 30% (Sharma et al. 2013). Similar findings were reported by Gupta et al. (2009) for *P. juliflora* with a total sugar yield of 586.6 mg/g biomass obtained from enzymatic saccharification of sulphuric acid pretreated delignified biomass at optimal conditions of 120 °C for 60 min with 3% H_2SO_4 and loaded with a mixture of 3.0 U of filter paper cellulase (FPase) and 9.0 U of β-glucosidase. The statistical analysis of total sugar yield obtained from hydrolysis of selected samples with different enzyme loadings indicated that there was a significant difference between the sugar yields obtained from the samples hydrolyzed. The effect of enzyme loading was significant ($p < 0.01$) on total sugar yield obtained from all the samples hydrolyzed at 1% level of significance.

The per cent carbohydrate conversion of different samples hydrolyzed with various enzyme loading levels ranged from 4.11 to 90.86. As the enzyme loading level increased, the per cent carbohydrate conversion also increased for both untreated and pretreated samples (Fig. 3).

The per cent carbohydrate conversion increased drastically with 15% enzyme loading and thereafter increased gradually with 30% enzyme loading for all the

Table 4 Total sugar yields of untreated and pretreated samples hydrolyzed with different enzyme loadings

Pretreatment	Total sugar yield (mg/g biomass)		
	Enzyme loading, % (g enzyme protein/g biomass)		
	0	15	30
Untreated	46.8 ± 2.4	233.2 ± 7.6	373.5 ± 8.7
120 °C, 60 min, 2% NaOH	33.2 ± 1.8	449.7 ± 8.2	583.9 ± 7.3
100 °C, 60 min, 2% NaOH	28.3 ± 2.1	378.3 ± 6.9	541.5 ± 4.3
140 °C, 30 min, 1% NaOH	26.4 ± 1.7	361.6 ± 5.4	514.4 ± 3.7

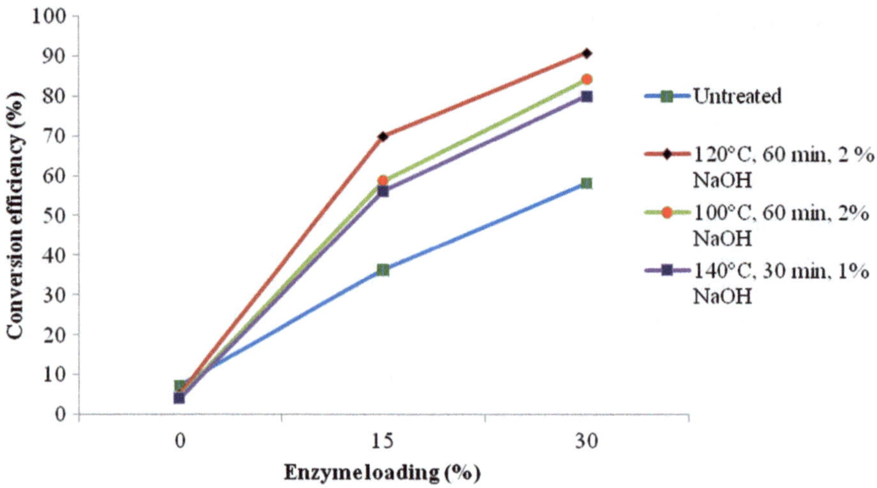

Fig. 3 Carbohydrate conversion of untreated and pretreated samples hydrolyzed with different enzyme loadings

samples. More than 70% of carbohydrates were converted with 15% enzyme loading. The maximum carbohydrate conversion of 90.86% was recorded for the sample pretreated at optimal conditions (120 °C, 60 min, 2% NaOH) which was loaded with 30% enzyme. Whereas the sample pretreated at 140 °C for 30 min with 1% NaOH recorded minimum carbohydrate conversion (4.11%) without enzyme loading. The maximum carbohydrate conversion (91.8%) was reported by Sharma et al. (2013) for KOH-pretreated switchgrass at 30% enzyme loading with Cellic$^{(R)}$ CTec2 cellulase enzyme (Novozymes, North America, Franklinton) at optimal pretreatment conditions were in close agreement with the maximum carbohydrate conversion (90.86%) obtained in this study for NaOH pretreated *P. juliflora* at 30% enzyme loading with CTec2 cellulase enzyme (Novozymes, Beizing, China). The statistical analysis of per cent carbohydrate conversion achieved from hydrolysis of selected samples with different enzyme loadings indicated that there was a significant difference between the per cent carbohydrate conversion obtained from the samples hydrolyzed. The effect of enzyme loading was significant ($p < 0.01$) on per cent carbohydrate conversion obtained from all the samples hydrolyzed at 1% level of significance.

However, the untreated sample without enzyme loading recorded higher carbohydrate conversion than all the pretreated samples with 0% enzyme loading. Similar trends were reported by Sharma et al. (2013). This may be due to the fact that only soaking effect could not convert all the carbohydrates of pretreated samples without enzyme loading which may be attributed to the loss of carbohydrates incurred during the pretreatment conditions. Based on maximum total sugar yield generated and highest carbohydrate conversion achieved, the sample pretreated at optimal conditions (120 °C, 60 min, 2% NaOH) and hydrolyzed with

30% enzyme loading was selected as optimum for subsequent fermentation for bioethanol production.

Fermentation Profile of Enzymatic Hydrolysate

The results pertaining to the batch fermentation of enzymatic hydrolyzates obtained from 30% enzyme loading on *P. juliflora* pretreated at 120 °C for 60 min with 2% NaOH concentration with *S. cerevisiae* are presented in Figs. 4 and 5.

As the fermentation time prolonged, ethanol yield increased regularly up to 24 h (21.84 g/l) and thereafter it declined. However, a regular decrease in sugars available in the sample was observed as the fermentation time increased. Further, it was observed that the ethanol productivity ranged from 0.39 to 1.14 g/l/h. As the fermentation time increased, drastic increase in ethanol productivity was observed up to 6 h, then it increased regularly up to 12 h and thereafter decrease in ethanol productivity was observed up to 36 h (Fig. 4).

The ethanol yield per g of fermentable sugars increased regularly up to 24 h (0.47 g/g) and thereafter it declined, as the fermentation time prolonged. Similar trend was observed for the ethanol yield per g of dry biomass (Fig. 5). Fermentation of enzymatic hydrolyzate containing 46.71 g sugar/l sample using *S. cerevisiae*, gave maximum ethanol of 21.84 g/l with yield (0.47 g/g sugar) and productivity (0.91 g/l/h) after 24 h of fermentation. The results are in close agreement with the

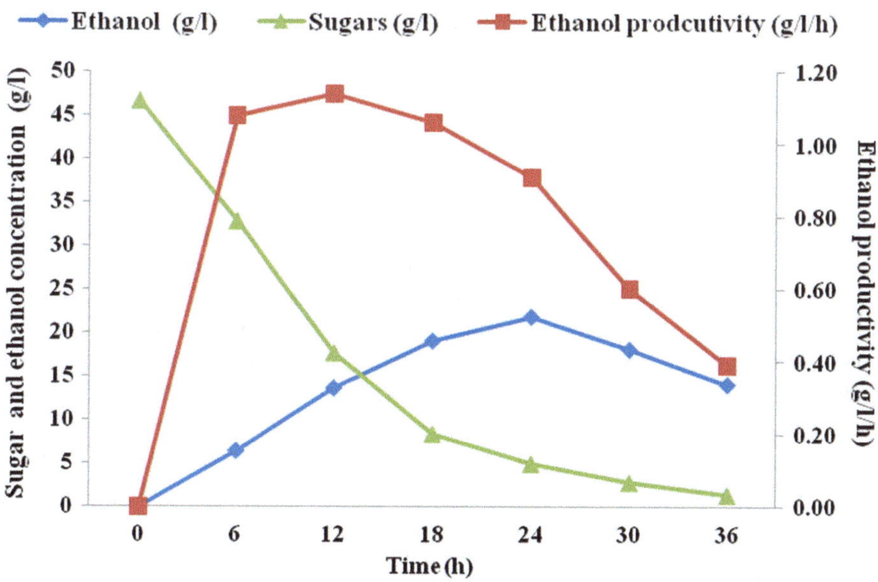

Fig. 4 Ethanol yield, sugar concentration and ethanol productivity during fermentation process

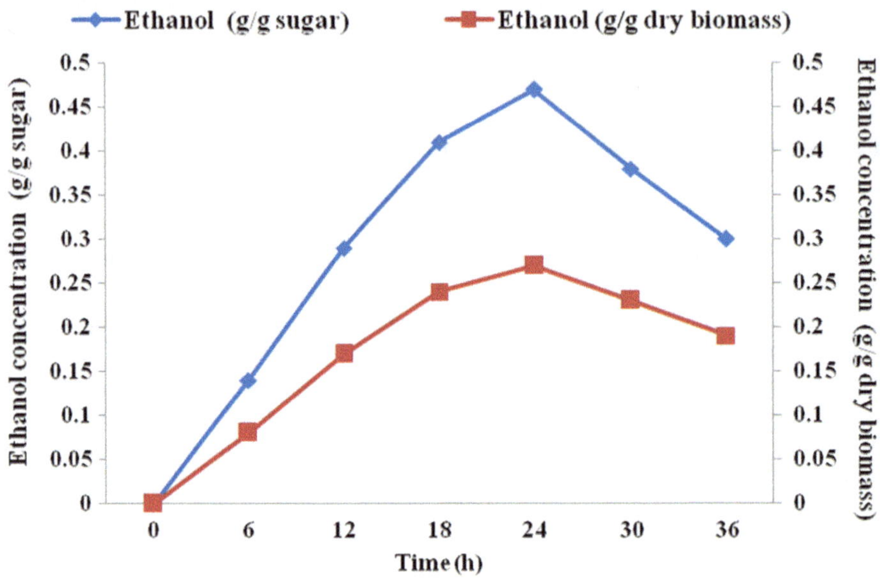

Fig. 5 Ethanol yield (g/g sugar) and ethanol yield (g/g dry biomass) during fermentation process

findings of experiment conducted by Gupta et al. (2009) on separate hydrolysis and fermentation (SHF) of *P. juliflora* for the production of cellulosic ethanol by *S. cerevisiae* with maximum ethanol (18.52 g/L) with yield (0.49 g/g sugar) and productivity (1.16 g/l/h) after 16 h.

Conclusions

Pretreatment of ground *P. juliflora* with NaOH resulted in promising lignin reduction of over 74% at 140 °C for 1 h with 3% NaOH concentration. Maximum lignin reductions at different temperatures were all obtained at the combinations of highest NaOH concentrations and longest treatment times, which indicated a close relationship between pretreatment severity and lignin reduction. Since, increasing pretreatment intensity does not necessarily lead to higher sugar recovery due to greater biomass solubilisation and lesser solids recovery, lignin reduction, though important, alone may not be an appropriate indicator for overall pretreatment effectiveness. The pretreatment combination of 120 °C, 60 min, 2.0% NaOH was chosen as the optimum for further enzymatic hydrolysis for fermentable sugar production as it had highest combined desirability. As the enzyme loading level increased, the per cent total sugar yield also increased for both untreated and pretreated samples. More than 70% of fermentable sugars were generated with 15%

enzyme loading. The sugar yield was maximum (583.9 mg/g biomass) with 30% enzyme loading at the end of 72 h of hydrolysis for the sample pretreated at optimal conditions (120 °C, 60 min, 2% NaOH) with a maximum carbohydrate conversion of 90.86%. A total ethanol of 270 kg per ton of dry *P. juliflora* could be produced by the fermentation of cellulosic hydrolyzates obtained from 30% enzyme loading on *P. juliflora* pretreated at 120 °C for 60 min with 2% NaOH concentration using *S. cerevisiae*.

Acknowledgements The authors would like to acknowledge the training imparted by Dr. Ratna Sharma-Shivappa, Associate Professor, Department of Biological and Agricultural Engineering, North Carolina State University, Raleigh, USAon "Fermentation Technology". Further, the authors would like to extend sincere thanks to the Novozymes, Beijing, China for sponsoring CTec2® Cellulase enzyme complex for conducting research on hydrolysis.

References

Alriksson B (2006) Ethanol from lignocellulose: alkali detoxification of dilute-acid spruce hydrolysates. Licentiate thesis, Karlstad University Studies, p 30

Bothast RJ, Saha BC (1997) Ethanol production from agricultural biomass substrates. Adv Appl Microbiol 44:261–286

Choge SK, Pasiecznik NM, Harvey M, Wright J, Awan SZ, Harris PJ (2007) *Prosopis pods* as human food, with special reference to Kenya. Nairobi Kenya For Res Inst 33(3):419–424. http://www.wrc.org.za

Eckard AD, Muthukumarappan K, Gibbons W (2012) Pretreatment of extruded corn stover with polyethylene glycol to enhance enzymatic hydrolysis: optimization, kinetics, and mechanism of action. Bioenergy Res 5:424–438

Eylen DV, Femke VD, Kabel M, Bont J (2011) Corn fiber, cobs and stover: enzyme-aided saccharification and co-fermentation after dilute acid pretreatment. Bioresour Technol 102:5995–6004

Ghose TK (1987) Measurement of cellulase activities. Pure Appl Chem 59:257–268

Gupta R, Sharma KK, Kuhad RC (2009) Separate hydrolysis and fermentation (SHF) of *Prosopis juliflora*, a woody substrate, for the production of cellulosic ethanol by *Saccharomyces cerevisiae* and *Pichia stipitis* -NCIM 3498. Bioresour Technol 100:1214–1220

Hopkins M (2007) Cooking up a smoky solution. Nature News, doi:10.1038/news070813-1, News

Kodaganti BP (2011) Simultaneous saccharification and fermentation of Arundo donax—comparison of feeding strategies. www.chemeng.lth.se/E655.pdf. Accessed 24 May 2012

Miller GL (1959) Use of dinitrosalicylic acid reagent for determination of reducing sugars. Anal Chem 31:426–428

Mosier N, Wyman C, Dale B, Elander R, Lee YY, Holtzapple M, Ladisch M (2005) Features of promising technologies for pretreatment of lignocellulosic biomass. Bioresour Technol 96 (6):673–686

Rajput SS, Tewari MC (1986) The role of prosopis in wasteland development. In: Patel VJ (ed) Javrajbhai Agroforestry Center, Surendrabag, Gujarat, India

Selig M, Weiss N, Ji Y (2008) Enzymatic saccharification of lignocellulosic biomass. Laboratory analytical procedure (LAP). National Renewable Energy Laboratory, Golden

Sharma R, Palled V, Sharma-Shivappa RR, Osborne Jason (2013) Potential of potassium hydroxide pretreatment of switchgrass for fermentable sugar production. Appl Biochem Biotechnol 169:761–772

Silverstein RA, Chen Y, Sharma-Shivappa RR, Boyette MD, Osborne J (2007) A comparison of chemical pretreatment methods for improving saccharification of cotton stalks. Bioresour Technol 98:3000–3011

Sluiter A, Hames B, Hyman D, Payne C, Ruiz R, Scarlata C, Sluiter J, Templeton D, Wolfe J (2005a) Determination of total solids in biomass and total dissolved solids in liquid process samples. Laboratory Analytical Procedure (LAP). National Renewable Energy Laboratory, Golden

Sluiter A, Hames B, Ruiz R, Scarlata C, Sluiter J, Templeton D, Crocker D (2008) Determination of structural carbohydrates and lignin in biomass. Laboratory Analytical Procedure (LAP). National Renewable Energy Laboratory, Golden

Sluiter A, Hames B, Ruiz R, Scarlata C, Sluiter J, Templeton D (2005b) Determination of ash in biomass. Laboratory Analytical Procedure (LAP). National Renewable Energy Laboratory, Golden

Sun Y, Cheng J (2002) Hydrolysis of lignocellulosic materials for ethanol production: a review. Bioresour Technol 83:1–11

Thuesombat P, Thanonkeo P, Laopaiboon L, Laopaiboon P, Yunchalard S, Kaewkannetra L, Thanonkeo S (2007) The batch ethanol fermentation of jerusalem artichoke using saccharomyces cerevisiae. J Sci Technol 7:93–96

Xu J, Cheng JJ, Sharma-Shivappa RR, Burns JC (2010) Sodium hydroxide pretreatment of switchgrass for ethanol production. Energy Fuels 24:2113–2119

Oil Extraction, Biodiesel Production and CI Engine Investigation Using *Madhuca indica* Methyl Ester

Amit Karwade, Girish Bhiogade, J.G. Suryawanshi and A.V. Bhujade

Abstract In recent years, the environmental concern and fossil fuel scarcity have led the government in many countries to formulate policy for alternative fuel development and emission control. The main purpose of this study was to try minor oil yielding varieties of nonedible oils like *Madhuca indica* oil to produce *Madhuca indica* methyl ester (MIME). The use of this type of biodiesel will definitely help to augment the gap in petroleum fuel demand and supply. In this direction Madhuca oil methyl ester has been tried as CI engine fuel. *Madhuca indica* methyl ester was made using alkaline, acidic catalyst and methanol with single and double stage transesterification process. The optimization of the transesterification process has been done to know the composition of chemicals required for the optimum yield of the ester. The properties of *Madhuca indica* methyl ester are found to be comparable to diesel fuel to great extent. The engine test was conducted to measure the engine performance and emission parameters of the fuel blends of MIME. The data obtained were tabulated and analyzed. The performance and emission characteristics of CI engine indicated a promising trend for MIME. The biodiesel economics reveals that the MIME and blends have good potential as CI engine fuel.

Introduction

The development of every country depends on utilization and production of energy and also utilization of natural resources. India is the fastest growing economy of the world, securing ninth position in the largest economy in the world. The share of petroleum products in the production of energy is 1%. In all petroleum products

A. Karwade · G. Bhiogade (✉) · J.G. Suryawanshi
Department of Mechanical Engineering, Visvesvaraya National Institute of Technology, Nagpur 440010, India
e-mail: gebhiogade@gmail.com

A.V. Bhujade
Department of Mechanical Engineering, Nagpur University, Nagpur 440033, India
e-mail: brauney1100@yahoo.com

V.P. Singh et al. (eds.), *Energy and Environment*, Water Science and Technology Library 80, https://doi.org/10.1007/978-981-10-5798-4_20

share of diesel is 41.63% maximum than others. 30% production of crude oil is done in India and remaining 70% requirement are met by imports. Per capita consumption of vegetable edible oil in India is 14.3 kg/annum and it is increasing 3–4% per year. India's domestic production of vegetable oil is decreasing and import is increasing year by year (Agarwal 2007; Demirbas 2007; Carraretto et al. 2004).

Madhuca found in India are *Madhuca indica* and *Madhuca longifolia*. It grows up to a height up to 70 ft. The tree matures and starts bearing 8–15 years, and fruits up to 60 years. The kernels are 70% of seed by weight and oil content is 52%. The trees have flowers as the season extends from February to April. The yield of 95% alcohol is 405 l from one ton of dried flowers. The *Madhuca indica* seeds were purchased from "Nature Conservation Society Amravati" the seeds were collected from tribal peoples of Melghat, Madhya Pradesh, India. The production of biodiesel from *Madhuca indica* oil having good scope in India and it will help to reduce the load on the mineral diesel consumption and production and therefore it will be helpful for the reducing the import of mineral diesel. *Madhuca indica* oil is nonedible oil source and biodiesel from *Madhuca indica* oil is environmentally friendly which produces fewer emissions as compared to diesel. We can extract 3.5 tons of oil from 10 tons of seeds by employing good technique of oil extraction that would cost Rs. 30 to Rs. 40 per liter of *Madhuca indica* oil (Ghadge and Raheman 2005; Sharma and Singh 2010; Goud et al. 2006).

The extracted crude *Madhuca indica* oil needs pre-treatment for achieving maximum yield of biodiesel. Extracted oil containing many impurities, such as suspended particles, sediments, pigments, soap, moisture, wax, free fatty acids, may affect the yield of biodiesel during the transesterification. It was necessary to improve these impurities before transesterification. Cloth filtration, caustic neutralization, and acid esterification processes were carried out before the transesterification process for refining the oil, lowering the value of free fatty acid. Followed by the neutralization acid esterification process was done. As the alcohol reacting with free fatty acid of oil it forms fatty acid methyl ester. FAME is suitable for transesterification for maximum yield of biodiesel 94–98% (Puhan et al. 2005; Padhi and Singh 2010).

In the transesterification process, FAME was heated and stirred with the mixture of alkyl catalyst NaOH 1% in w/w of oil and methanol with 6:1 molar ratio with oil. The mixture was heated at 60–65 °C with electric heater and stirred 100–200 RPM with magnetic stirrer for 2–3 h. Glycerol and ester were separated in the separating funnel. The physiochemical properties were tested and observed that, biodiesel having similar properties like diesel except flash point which is 182 °C for biodiesel. High viscosity of biodiesel can cause fuel injector chocking and fuel filter clogging.

The performance and emission test of diesel and biodiesel with various blends was performed. The performance parameters were speed, torque, power and thermal efficiency. Emission parameters were CO_2, CO, HC, NOx and smoke density. Performance was done and the result was plotted by using diesel, biodiesel and its blends of 25, 50 and 75%.

Fig. 1 Oil expeller machine (VNIT Nagpur)

Oil Extraction Process

The oil expeller method is adopted for *Madhuca indica* oil extraction. The expeller make is Rajlaxmi Engineering Corporation, India and it is driven by Field Marshal FMS 12 CI engine. Vertical steam boiler used for heating and moistening the oil seeds, so as to reduce the crushing pressure required for crushing the seeds (Fig. 1).

Oil seeds were carried out in the kettle of oil expeller, where it were moistened and heated by supplying the steam through a pipe from vertical steam boiler at 110 °C. The heating and moistening were necessary to reduce crushing pressure and avoiding jamming of brass screw. The oil seeds were primarily crushed in the kettle by rotating metallic strip coupled with shaft driven by the engine through the pulley and belt drive and bevel gear arrangement.

Biodiesel Production

The extracted oil containing impurities like suspended particles, pigments moisture, free fatty acid, soap, wax. These factors create problems for biodiesel production with direct use of the oil in transesterification. It was necessary to remove these impurities from the oil for maximum yield of biodiesel from the *Madhuca indica* oil. Good pre-treated oil yields 94–98% of biodiesel. The necessary oil pre-treatments were carried out.

Cloth Filtering

Cloth filtering of crude oil was carried out for removing suspended particles, sediments, wax and fats. The extracted oil had high viscosity and pouring point hence the oil was heated at 60–70 °C and then the oil was poured into the bag of cloth filter. Filtered oil came out was then collected into the bucket. The oil appeared in greenish yellow color. The impurity removed by the filtration was 2.1 kg from 14 kg of crude oil at 15% loss.

Caustic Pre-treatment

Followed by filtering process, caustic pre-treatment was done for removing a saponification value of *Madhuca indica* oil. *Madhuca indica* oil has saponification value up to 187–197 that results in no yield of biodiesel and higher amount of soap formation.

One liter of Madhuca oil carried out in a round bottom flask was heated to 65 °C. The NaOH flakes of 10% by weight that is 100 g dissolved in 50 g of water and the mixture heated at 65 °C and then the dissolve mixture added to the oil. The mixture was stirred vigorously (250 rpm) so as to break emulsions that forming during the process. The reaction was carried out at 65 °C for 60 min to break any emulsion that might have formed during neutralization. For removing the soap formation, water washing was done. The soap settlement requires 24 h. In the separating funnel bottom layer soap and water (white color) and at the upper layer neutralized oil (yellow color) was obtained. The neutralized oil, then heated for 110 °C to remove the moisture from the oil. The oil obtained from the naturalization process was 750 ml in 1 l of oil (Fig. 2).

Acid Pre-treatment

Followed by caustic neutralization, the acid esterification process was carried out for achieving maximum yield of biodiesel. One liter filtered Madhuca oil was heated to 110 °C in rounded bottomed flask so as to remove the moisture present in it. After achieving this temperature, the heated oil appeared in red color and was cooled to 60 °C. The mixture of sulphuric acid (1% based on oil weight, approx. 5–10 ml) in 0.15 w/v methanol (150 ml) was added to the reaction flask preheated oil. The mixture was heated to 60 °C and stirred by a magnetic stirrer at 100–150 RPM for 6 h. And then the mixture was separated in separating funnel where the upper layer was unreacted methanol, middle layer was fatty acid methyl ester in reddish color and at lower layer fats were found (Acharya et al. 2011; Jindal et al. 2010; Suryawanshi et al. 2013) (Fig. 3).

Fig. 2 Separation of oil and soap

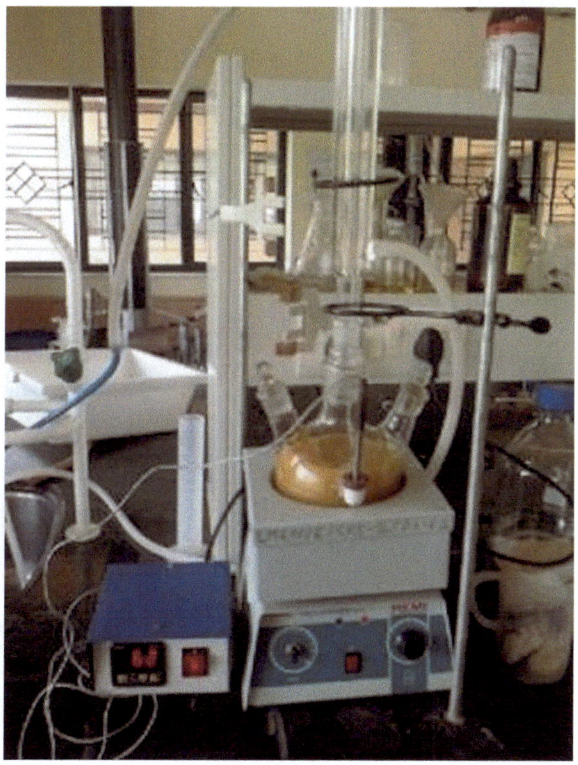

Fig. 3 Acid esterification process

Transesterification Process

In transesterification using NaOH as catalyst, the Madhuca oil was preheated at 110 °C for 20–25 min and mixed with methanol–NaOH solution. The transesterification was carried out using different % of NaOH varying from 0.5 to 1% (10 g NaOH per liter of oil) of oil weight and methanol 30% of oil weight in 1:6 molar ratios. The reaction mixture was then heated to 650–750 °C at a low magnetic stirrer speed of revolution because at high temperature and high speed, mixture produces more amount of soap formation and due to this the separation of MIME becomes more difficult. Hence the speed of magnetic stirrer was 100–150 RPM for 120–150 min. The reaction mixture was to be allowed to settle overnight in separating funnel to separate methyl ester upper layer and glycerine at lower layer (Fig. 4).

Normally the transesterification reaction requires 3 mol of alcohol for one mol of triglycerides to three mole of fatty acid ester and one mol of glycerol. Excess amount of alcohol increases conversion of fats into esters within a short time. So the yield of biodiesel increases with increase in the concentration of alcohol up to a

Fig. 4 Separation of glycerol

certain concentration. However, a further increase in alcohol content does not increase the yield of biodiesel, but it also increases the cost of alcohol recovery.

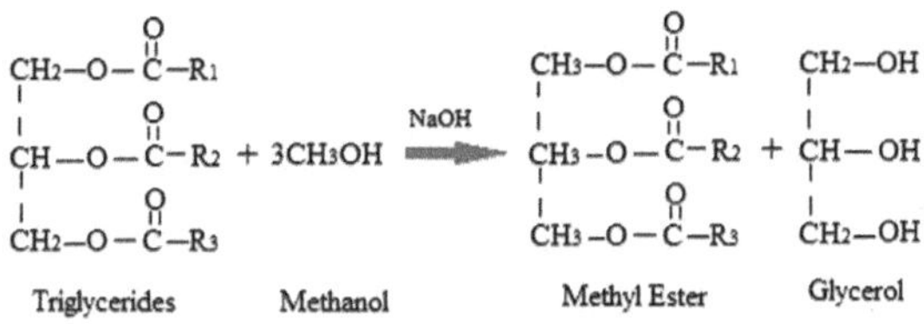

The filtration loss was 15%, loss in neutralization 25% and loss in transesterification process was 2%. Therefore 625 ml of biodiesel could obtain in 1000 ml of extracted *Madhuca indica* oil (Hideki et al. 2001; Ramadhas et al. 2005; Meher et al. 2004; Agarwal and Das 2001).

Comparison of Properties of Diesel and Biodiesel

Table 1 shows the properties of diesel, *Madhuca indica* oil and *Madhuca indica* Methyl Ester (MIME).

Higher Flash Point in MIME than diesel indicates ease of handling and transportation but higher pour point and cloud point in MIME create problem for cold start conditions (Yuan et al. 2005).

Results and Discussion

Experiments were initially carried out on the engine at all loads to create baseline data. The engine was stabilized before taking all the readings. Blends of different proportions of *Madhuca indica* Methyl Ester (MIME) and Diesel ranging from 25 to 100% were used to run this single cylinder engine. Table 2 shows the engine specification.

Figure 5 shows a characteristic curve between load and brake thermal efficiency of biodiesel with various blends and diesel. B50 has a lowest brake thermal efficiency at medium to heavy loads. Maximum brake thermal efficiencies achieved

Table 1 Properties of diesel, *Madhuca indica* oil and *Madhuca indica* methyl ester

Property	Diesel	M.I. oil	MIME
Density (kg/m^3)	835	945	872
Kinematic viscosity (cSt)	2.4	2	4.0
Flash point (°C)	70	226	182
Fire point (°C)	76	250	204
Calorific value (MJ/kg)	43	3	39.36
Cetane number	47	NM	50

Table 2 Engine specifications

Engine	Parameters
Make	Kirloskar TV1
Bore × stroke	87.5 mm × 110 mm
Cubic capacity	661 cm^3
Compression ratio	17.5:1
Rated output	5.2 kW @1500 rpm
Fuel injector pressure	210 bar
Fuel injection type	Mechanical

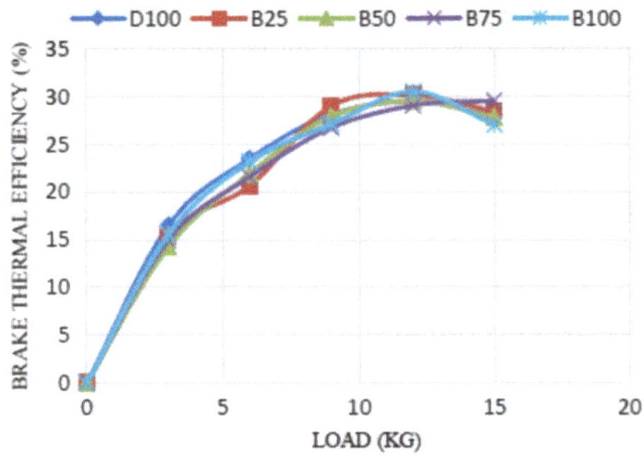

Fig. 5 Variation of brake thermal efficiency versus load

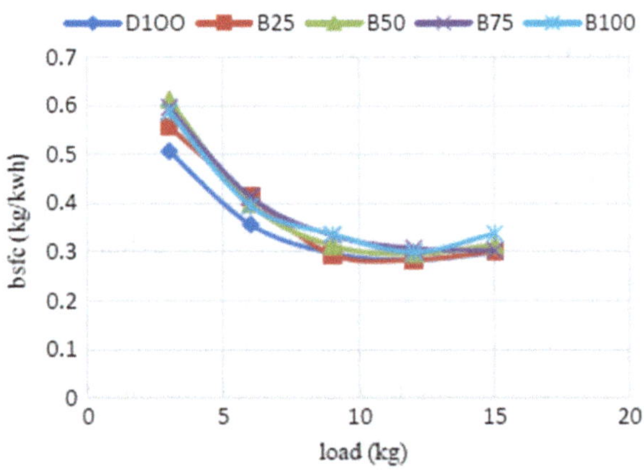

Fig. 6 Variation of BSFC versus load

by B100 at 12 kg load was 30.54% with brake power 3.615 kW followed by B25 at 12 kg load 30.16% produced 3.572 kW brake power. Use of B75 will be more efficient at higher loads than biodiesel, diesel and other blends.

Figure 6 shows the variation of brake specific fuel consumption with load. Brake specific fuel consumption remains high at idling and low load condition as the piston has to work over gas pressure inside the cylinder, the inertia of rotating parts and friction between the cylinder wall and piston. At low load, low brake power BSFC of B50 and B100 was highest, BSFC for diesel and B25 were lowest and B75 was in the middle position in case of fuel consumption. At medium loads BSFC for B25 was the lowest for all blends, followed by diesel and B75.

Figure 7 shows the variation of smoke opacity with loads. It has been found that smoke opacity of B75 and B100 was higher at low and medium loads, but it was gone decreasing with increase in load. Smoke opacity of B25 and B50 was lower at low and medium loads, but was increasing with increase in loads. D100 has had the highest smoke opacity than other blend at high load.

Figure 8 shows emissions curve between load and CO, the CO emissions are decreasing with an increase in the load. B50 emits maximum CO emissions than all blends, CO emissions of B100 were lesser than all blends and diesel at low loads, but was higher than B25 at high load. CO emissions from diesel were more at high load because of rich air–fuel mixture and insufficiency of oxygen to convert CO completely to CO_2.

Fig. 7 Variation of smoke opacity versus load

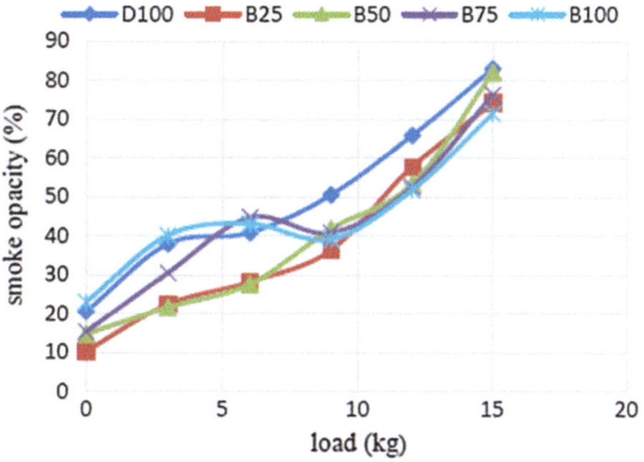

Fig. 8 Variation of CO versus load

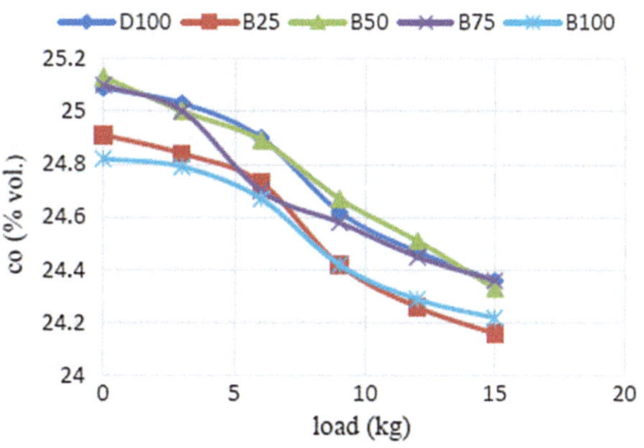

Fig. 9 Variation of HC versus load

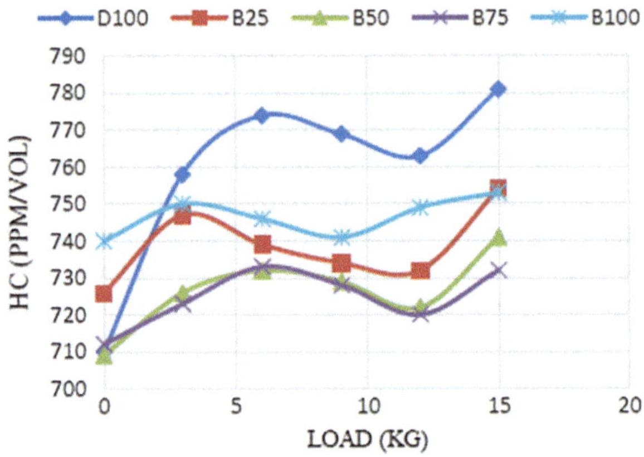

Fig. 10 Variation of NOx versus load

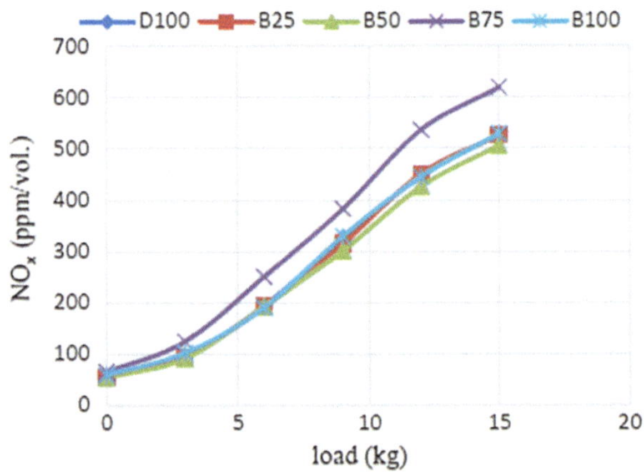

Figure 9 represents the relation between load and unburnt hydrocarbon emissions. HC emissions of B100 were higher than that of all blends and diesel at low load. Reason behind that the viscosity of biodiesel is high and because of that dribbling of fuel near the sac volume of fuel injection may lead to cause unburnt HC emission from exhaust. For medium to higher loads HC formations in biodiesel and B75 start decreasing as the oxidation of HC into CO_2 and H_2O at high temperature takes place due to the presence of oxygen in the fuel. Unburnt HC emissions for D100 were the lowest than that of all blends and B100 due to higher calorific value leads to better combustion of fuel at low load. But at high load, HC emissions were more because of richness of air fuel mixture and insufficiency of oxygen for oxidation of HC.

Figure 10 shows variation between load and NOx emission characteristics. The nitrogen present in the supplied air remains inert at low temperature, but at high temperature, it reacts with the oxygen of air fuel mixture and forms oxides of nitrogen (NOx). B75 produced higher NOx as the presence of oxygen in fuel and higher at high load because of high temperature inside the cylinder and higher calorific value than B100. NOx emission for B100 was more than D100, B25 and B50 at low to medium loads. NOx emission of B50 was lower than all blends at all load conditions.

Conclusion

The work presented in this paper was the extraction of oil from *Madhuca indica* seeds for the production MIME. Properties of MIME with the standard procedure were tested, compared with mineral diesel and performance and emission test was carried on CI Engine.

Oil was extracted economically from oil seeds by using oil expeller. From 68 kg of oil seed, oil extracted was 14 kg. Extracting the impurities, 1 l oil was obtained from 4.86 kg of *Madhuca indica* oil seeds. Filtration, neutralization and acid esterification process are required for maximum yield of biodiesel from transesterification process.

Smoke opacity was lesser for B25 and B50 blends than diesel at all loading. Smoke opacity for B100 and B75 decreases further than diesel at high load. HC emissions of B50 and B25 were observed less at medium to high loads. At low loads HC emissions of biodiesel were greater than diesel because of high viscosity. At high loads HC emission in diesel is quite higher compared to biodiesel because of richness of air fuel mixture and insufficiency of oxygen for oxidation of HC.

B75 produced higher NOx as the presence of oxygen in fuel and higher at high load because of high temperature inside the cylinder and higher calorific value than B100. NOx emission for B100 was more than D100, B25 and B50 at low to medium loads. NOx emission of B50 was lower than all blends at all load conditions.

As per the performance and emissions study of CI engine it can be concluded that the use of MME with B50 will be a potential alternative of CI engine.

References

Acharya SK et al (2011) Performance analysis of karanja and kusum oils as alternative bio-diesel fuel in diesel engine. Int J Agric Biol Eng 4(2):23–28

Agarwal AK (2007) Bio-fuels applications as fuels for internal combustion engines. Prog Energy Combust Sci 33(3):233–271 (1267)

Agarwal AK, Das LM (2001) Biodiesel development and characterization for use as a fuel in compression ignition engines. J Eng Gas Turbine Power 123:440–447

Carraretto C et al (2004) Biodiesel as alternative fuel: experimental analysis and energetic evaluations. Energy 29:2195–2211

Demirbas A (2007) Progress and recent trends in Bio-fuels. Prog Energy Combust Sci 33

Ghadge SV, Raheman H (2005) Biodiesel production from mahua (*Madhuca indica*) oil having high free fatty acids. Biomass Bioenergy 28:601–605

Goud VV, Patwardhan AV, Pradhan NC (2006) Studies on the epoxidation of mahua oil (*Madhumica indica*) by hydrogen peroxide. Bioresour Technol 97:1365–1371

Hideki F, Akihiko K, Hideo N (2001) Biodiesel fuel production by transesterification of oils. J Biosci Bioeng 92(5):405–416

Jindal S et al (2010) Experimental investigation of the effect of compression ratio and injection pressure in a direct injection diesel engine running on Jatropha methyl ester. Appl Therm Eng 30:442–448

Meher LC, Naik SN, Das LM (2004) Methanolysis of *Pongamia pinnata* (Karanja) oil for production of biodiesel. J Sci Ind Res 63:913–918

Padhi SK, Singh RK (2010) Optimization of esterification and transesterification of Mahua (*Madhuca indica*) oil for production of biodiesel. J Chem Pharm Res 599–602. ISSN: 0975–7384

Puhan S, Vedaraman N, Ram BVB (2005) Mahua oil (*Madhuca indica* oil) methyl ester as biodiesel-preparation and emission characteristics. Biomass Bioenergy 87–93

Ramadhas AS, Jayaraj S, Muraleedharan C (2005) Biodiesel production from high FFA rubber seed oil. Fuel 84:335–340

Sharma YC, Singh B (2010) A hybrid feed stock for a very efficient preparation of biodiesel. Fuel Process Technol 91:1267–1273

Suryawanshi JG, Bhiogade GE, Paikra AS (2013) Performance and emission analysis of *Schleichera oleosa* oil methyl ester on CI engine. In: 23 NCICEC, SV National Institute of Technology, Surat, India, vol 336, pp 1–7

Yuan W, Hansen AC, Zhang Q (2005) Vapor pressure and normal boiling point predictions for pure methyl esters and biodiesel fuels. Fuel 84:943–950

Design and Development of Producer Gas-Based Heat Exchanger for Drying Application

D.K. Vyas, S.S. Kapdi and M.L. Gaur

Abstract Gasification is a thermochemical process, which converts solid biomass in oxygen-deficient environment into combustible producer gas. The products of gasification are used for thermal applications, electricity generation of combined thermal and electrical power outputs (cogeneration). The updraft gasifier reactor was designed for 4–10 kg/h fuel consumption rates with the thermal capacity of 10,000–25,000 kcal/h. Combustion chamber of producer gas and shell and tube type heat exchanger was designed and developed. The performance of the designed updraft gasifier-based heat exchanger was evaluated at different fuel consumption rates (4, 6, 8 and 10 kg/h) and air flow rates (200, 300, 400, and 500 m^3/h) using various biomass fuels. Diameter of reactor, height of reactor, area required, and volume of reactor for the updraft biomass gasifier were 0.40 m, 1.00 m, 0.1256 m^2, and 0.1256 m^3, respectively. The amount of carbon monoxide (CO), hydrogen (H_2), methane (CH_4), and calorific value of producer gas were decreased with the increase in fuel consumption rates. The maximum and minimum gasifier efficiency were found as 79.24, 80.11, and 69.18% and 75.88, 73.97, and 62.43% for maize cobs, *Prosopis juliflora*, and saw dust briquettes, respectively. The maximum heat exchanger effectiveness was found as 70.45, 74.88, and 74.73% at 4 kg/h fuel consumption rate and 500 m^3/h air flow rate using maize cobs, *P. juliflora*, and saw dust briquettes, respectively. The maximum hot air temperature 89, 92, and 83 °C was measured at 10 kg/h fuel consumption rate and 200 m^3/h air flow rate using all three fuels. The biomass gasification-based heat exchanger system is suitable for thermal application.

D.K. Vyas (✉)
Department of Renewable Energy, College of Agricultural Engineering and Technology, Anand Agricultural University, Godhra, India
e-mail: dhamo810@yahoo.com; dhamo810.caet@aau.in

S.S. Kapdi
Department of Bio Energy, College of Food Processing Technology & Bio Energy, Anand Agricultural University, Anand, India

M.L. Gaur
College of Agricultural Engineering and Technology, Anand Agricultural University, Godhra, India

V.P. Singh et al. (eds.), *Energy and Environment*, Water Science and Technology Library 80, https://doi.org/10.1007/978-981-10-5798-4_21

Keywords Biomass · Gasifier · Gasifier efficiency · Heat exchanger effectiveness

Introduction

Energy is a key input for technological, industrial, social, and economical development of a nation. A large number of consumers in domestic, agricultural, commercial, and industrial sectors are faced with a situation of energy availability that is characterized by inadequate quantity, poor quality, unaffordability, unsustainability, and negative environmental consequences. The challenge for the country is ensuring affordable clean energy for all in a sustainable manner. Agriculture and energy have always been tied by close links, but the nature and strength of the relationship have changed over time. The link between agriculture and energy output markets weakened in the twentieth century as fossil fuels gained prominence in the transport sector. The use of renewable resources would contribute to a country's economic growth, especially in developing countries, many of which have abundant biomass and agricultural resources that provide the potential for achieving self-sufficiency in materials. Biomass is an important type of renewable energy fuel source for electrical or thermal power production. Biomass includes agricultural residues, urban wastes even sewage sludge waste and it contributes a significant share of global primary energy consumption and its importance is likely to increase in future world energy scenarios (Vasudevan et al. 2005). Biomass used as fuel for energy production in heating or electricity generating applications, biomass is the fourth primary energy sources after coal, oil, and natural gas accounting for about 14% of the world's total energy supply.

There are three methods to convert biomass material into a useful form of energy, viz. the direct combustion, the biological conversion, and the thermochemical conversion (Zainal 1996). Gasification is a thermochemical process, which converts the solid biomass in oxygen-deficient environment into combustible producer gas. The products of gasification are used for thermal applications, electrical generation, or combined thermal and electrical power outputs (cogeneration). For thermal applications, the drying process is one of the biggest challenges for many industries to get a cheap and clean heat source; for example, timber drying and food processing. The major combustible components of producer gas are carbon monoxide, hydrogen, methane, and hydrocarbons. Organic byproducts are readily available in agricultural and agro-industrial sectors (Vyas 2003). The shell and tube type heat exchanger are closed coupled with the updraft gasifier for hot air generation for drying of the agricultural as well as industrial products. The air in the shell side passes across the vertically placed tubes and the hot flue gases in the tube side with two passes. The hot gases from the combustion chamber of producer gas will be induced up and down through the tubes of the heat exchanger using a draft fan fixed in the exhaust of the chamber. Gas-to-air heat exchanger will be used to heat up clean air as a thermal power output for drying process, space heating, or any other thermal process.

The small, rural-forested communities have the economic need to develop a wood products industry to replace the loss of large saw mills and maintain forest health. The potential of using producer (wood) gas to fire a hot water boiler for a small dry kiln capable of drying both softwood and hardwood lumber (Bergman 2005). The externally fired "turbo charger based" micro-gas turbine using biomass fuel in a small-scale flexible unit which could be used in the simple form without the electrical generator to fulfill the industrial demands of a cheap and clean hot air for different drying processes (Alattab and Zainal 2006). The gross efficiency of the micro-gas turbine depends on output power but not on the gas heating value, within our measurement accuracy of a gas engine (Rabou et al. 2007). The Olive mill technology generates a variety of biomass wastes: olive pits/stones and remaining pomace resultant from olive oil extraction. Thermodynamic calculations evaluate the optimum operating parameters of a small-scale gasification system. Simulation results showed the most important operating parameters are turbine inlet temperature (TIT), pressure ratio (PR) and hot side temperature difference of the heat exchanger (HTHE). High electric efficiency (20%) and overall efficiency (65%) were achievable with such a system (Vera et al. 2011).

A biomass gasifier coupled with the combustion chamber for heating air, which can be used for the thermal conditioning of poultry houses, drying of agricultural products, etc. The gasifier was capable of generating synthesis gas using firewood with different dimensions that, burned in a close coupled combustion chamber, efficiently generates clean warm air that can be used for agricultural purposes, such as space heating for broilers or grain drying (Silva et al. 2014).

Drying is a major heat energy consuming operation in agro processing industries. The potential problems associated with solar drying such as the inability of commercial and community producers to process agricultural and fishery products during inclement weather conditions and night time, a biomass fuelled energy source integrated to the dryer as a separate component will solve the weather dependent conditions. The heat energy obtained from combustion of biomass can be used directly or supplied through carriers like steam and hot air. With the development of technology for drying of food and agricultural products supply of hot air at reasonable cost and controlled temperatures and in regulated quantity has become extremely important. The design and development of a producer gas-based heat exchanger to generate hot air for drying application is discussed in this paper.

Methodology

Physical Properties and Proximate Analysis of Different Biomass

Available biomass namely maize cobs, *Prosopis juliflora*, and biomass briquettes considered for gasification material. The physical properties (i.e., size, bulk, and true density) and proximate analysis (i.e., moisture content, volatile matter, ash content, and fixed carbon) were determined using ASAE (ASTM 1983) and ASTM (ASABE 2006) standards for maize cob, *P. juliflora*, and biomass briquettes as fuel used in updraft gasifier. The calorific value of biomass was measured using advanced bomb calorimeter.

Performance Evaluation of Updraft Gasifier

The updraft gasifier is designed and developed at Department of Renewable Energy Engineering, College of Agricultural Engineering & Technology (CAET), Anand Agricultural University, Godhra. The gasifier has a single-stage chamber namely, gasifier-gasification chamber. This part is made of a mild steel sheet. It is a batch-feeding gasifier system, biomass feeding port is placed on the top of the gasifier chamber to feed the biomass into the gasifier and it is fit by clamp and isolated cover. Air is supplied with control valve to the gasification chamber at almost atmospheric pressure. The flow of air is passing through a bottom of the grate to provide a good mixing with the gases from the gasifier chamber. An incomplete combustion in the gasification chamber creates a very hot zone which pushes the gases upward and creates a stack effect which creates a vacuum pressure in the gasifier chamber and pulls the air through a controlled air intake in the down side and into the gasifier chamber to complete the gasification process. The schematic diagram of the experimental setup of the biomass updraft gasifier is shown in Fig. 1.

Development of Single Pass Shell and Tube Type Heat Exchanger

The high-temperature heat exchanger (HTHE) is the key to the success in the externally fired gas turbine (EFGT). It is required to transfer heat from the heat source like combustion chamber or furnace to the working fluid for the thermal application. At the higher temperature, the heat exchanger can provide the higher system efficiency (Perry 1963). The high-temperature heat exchanger had been designed and the design was based on single pass shell and tube type heat exchanger. The combustion gas temperature is assumed to be in the range of 500–1000 °C and the air temperature after the heat exchanger is assumed to be in the range of 60–240 °C. Different size air flow rates of the air blower were used to study the performance of the biomass gasifier-based heat exchanger for different fuels and fuel consumption rates. The shell and tube type heat exchanger for thermal application was thermally designed by trial and error calculations using the Kern method (Kern 1965).

Fig. 1 Actual photographic view of updraft gasifier

Heat transfer area (A) required:

$$A = \left(\frac{Q}{U_{o,\text{assm}} \times \text{LMTD} \times F_T}\right) \tag{1}$$

where

A	Heat transfer area, m^2
Q	Heat duty of the exchanger, kcal/s
$U_{o,\text{assm}}$	Overall heat transfer coefficient, kcal/h m^2 °C
LMTD	Log Mean Temperature difference, °C
F_T	Correction factor

$$\text{LMTD} = \left(\frac{(T_2 - T_1) - (t_2 - t_1)}{\ln((T_2 - T_1)/(t_2 - t_1))}\right) \tag{2}$$

$$F_T = (A_1/A_2)$$

$$A_1 = \left(\frac{(\sqrt{R^2+1}) \times \ln(1-S)}{1-RS}\right)$$

$$A_2 = (R-1) \times \ln\left(\frac{(2-S) \times \left(R+1-\sqrt{R^2+1}\right)}{(2-S) \times \left(R+1+\sqrt{R^2+1}\right)}\right)$$

$$R = \left(\frac{(T_1 - T_2)}{(t_2 - t_1)}\right)$$

$$S = \left(\frac{(t_2 - t_1)}{(T_1 - t_1)}\right)$$

Number of tubes (n_t) required to provide the heat transfer area (A):

$$n_t = \left(\frac{A}{\pi \times d_o \times L}\right) \quad (3)$$

where

n_t Number of tubes
d_i Inside tube diameter, m
d_o Outside tube diameter, m
L Tube length, m

The thermal performance of heat exchanger, effectiveness of the heat exchanger (ε) was calculated from Eq. 4:

$$\varepsilon = \left(\frac{C_{air}\left(T_{air\,out} - T_{air\,in}\right)}{C_{min}\left(T_{gas\,in} - T_{air\,in}\right)}\right) \times 100 \quad (4)$$

where

C_{air} Heat capacity rate for air, kcal/kg °C
C_{gas} Heat capacity rate for gas, kcal/kg °C

C_{min} is the minimum heat capacity rate and it is equal to C_{air} in this design, it was calculated from Eq. 5:

$$C_{air} = \dot{m}x\,C_p \quad (5)$$

To calculate the efficiency of the heat exchanger, the mass flow rate of the hot combustion gases were calculated from Eq. 6:

$$C_{\text{air}} \times (\Delta T_{\text{air}}) = C_{\text{gas}} \times (\Delta T_{\text{gas}}) \tag{6}$$

Total thermal power of the combustion gases ($Q_{\text{combustion gases}}$) was calculated from Eq. 7:

$$Q_{\text{combustion gases}} = C_{\text{gas}} \times (T_{\text{gas in}} - T_{\text{ambient}}) \tag{7}$$

Thermal power transferred to the hot air (Q_{air}) through heat exchanger was calculated from Eq. 8:

$$Q_{\text{air}} = C_{\text{air}} \times (\Delta T_{\text{air}}) \tag{8}$$

Efficiency of the heat exchanger (η_{HE}) was calculated from Eq. 9:

$$\eta_{\text{HE}} = [Q_{\text{air}}/Q_{\text{combustion}})] \times 100 \tag{9}$$

In order to understand the main factors affecting the heat exchanger efficiency (ε), Eq. 4 was simplified into Eq. 10:

$$\varepsilon = [(T_{\text{gas in}} - T_{\text{gas out}})/(T_{\text{gas in}} - T_{\text{ambient}})] \times 100 \tag{10}$$

The detailed design drawing and photographic view of the shell and tube type heat exchanger are shown in Figs. 2 and 3, respectively.

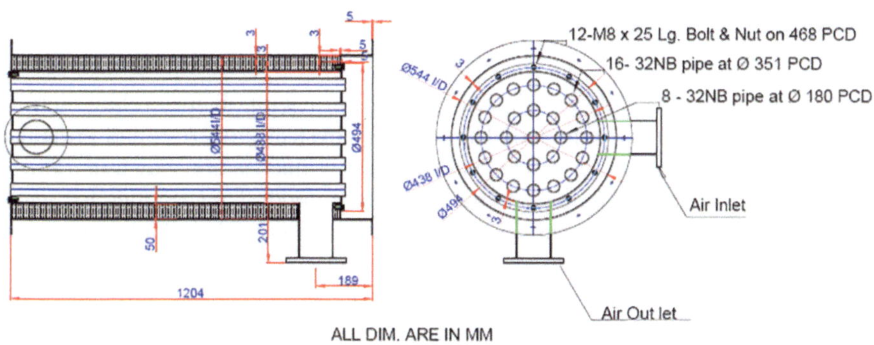

Fig. 2 Detail drawing of shell and tube type heat exchanger

Fig. 3 Actual photographic view of shell and tube type heat exchanger

Results and Discussion

Physical Properties and Proximate Analysis of Different Biomass

Physical properties of different biomass in terms of length, diameter, bulk density, and true density were determined. The average physical properties and proximate analysis of maize cobs, *P. juliflora*, and saw dust briquettes are shown in Table 1. The calorific value of different biomass indicated good characteristics for gasification because higher heat generated during combustion leads to high temperature in gasification, i.e., reaction zone.

Design Parameters of Updraft Biomass Gasifier and Shell and Tube Type Heat Exchanger

Table 2 shows the data on diameter of reactor, height of reactor, fuel consumption rate, specific gasification rate, volume of reactor, diameter and height of heat

Table 1 Average physical properties and proximate analysis of different biomass

S. No.	Physical properties (average)	Name of the biomass		
		Maize cobs	*Prosopis juliflora*	Saw dust briquettes
1	Length (mm)	89.41	40.23	25.33
2	Diameter (mm)	22.79	28.44	88.42
3	Bulk density (kg/m^3)	297.88	410.93	402.67
4	True Density (gm/cm^3)	0.33	0.81	0.77
5	Moisture content (%d.b.)	6.02	6.44	8.04
6	Ash content (%d.b.)	1.69	1.62	12.04
7	Volatile matter (%d.b.)	86.59	87.08	82.48
8	Fixed carbon (%d.b.)	11.72	11.30	5.48
9	Calorific value (kcal/kg)	3815	4199	4217

Table 2 Design parameter of developed updraft biomass gasifier-based combustor

S. No.	Design parameter	Dimensions
Updraft gasifier		
1	Diameter of reactor (m)	0.40
2	Height of reactor (m)	1.00
3	Area required (m^2)	0.1256
4	Volume of reactor (m^3)	0.1256
5	Fuel consumption rate (kg/h)	10.0
6	Specific gasification rate (kg/m^2 h)	79.62
7	Air flow rate (m/s)	0.024
8	Superficial velocity (m/s)	0.190
Shell and tube type heat exchanger		
9	Diameter of shell (m)	0.433
10	Height of shell (m)	1.204
11	Number of tubes	25
12	Inside diameter of tubes (m)	0.032
13	Outside diameter of tubes (m)	0.040

exchanger, number of tubes, inside and outside diameter of the tube designed, and development for experimentation.

Effect of Fuel Consumption Rate on Producer Gas Composition and Calorific Value for Various Biomass

The producer gas composition at different fuel consumption rates, i.e., 4, 6, 8, and 10 kg/h using different biomass is shown in Fig. 4. The value of CO, H_2, and CH_4 was decreased with increase of fuel consumption rates. It could be seen from the

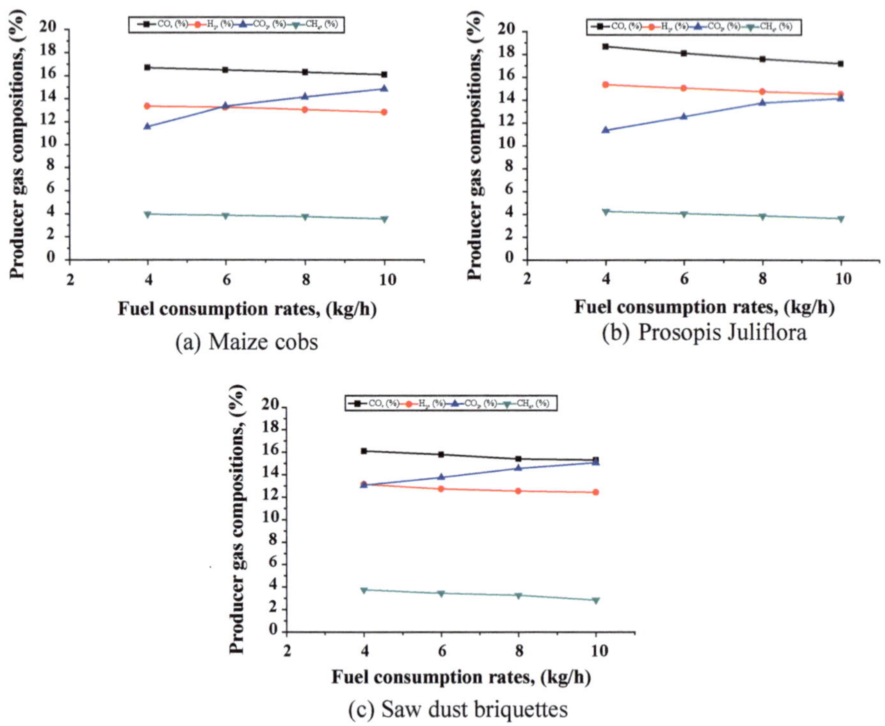

Fig. 4 Variation in producer gas compositions with fuel consumption rates using different biomass

statistical analysis that the producer gas composition was found statistically significant on 5% level of significance.

The calorific value of producer gas and efficiency of gasifier at different fuel consumption rates, i.e., 4, 6, 8, and 10 kg/h using different biomass are given in Table 3. The gasifier efficiency was decreased with increase in fuel consumption rate due to decrease in calorific value of the producer gas.

Effect of Air Flow Rate, Fuel Consumption Rate, and Gas Flow Rate on Air Outlet Temperature of Heat Exchanger

Variations in air outlet temperature of the heat exchanger at different fuel consumption rates and air flow rates using different biomass are tabulated in Table 4. It can be seen that value of air outlet temperature of the heat exchanger was increased with increase in fuel consumption rate but decreased with increase in air flow rates. The effect of interaction of these factors on air outlet temperatures was found statistically significant on 5% level of significance.

Table 3 Variation in calorific values of producer gas using different biomass at different fuel consumption rates

S. No.	Fuel consumption rates (kg/h)	Calorific value of producer gas (kcal/m^3)			Gasifier efficiency (%)		
		Maize cobs	*Prosopis juliflora*	Saw dust briquettes	Maize cobs	*Prosopis juliflora*	Saw dust briquettes
1	4.0	1201.46	1337.00	1159.48	79.24	80.11	69.18
2	6.0	1183.33	1292.60	1111.21	78.74	78.15	66.90
3	8.0	1162.77	1251.05	1074.93	77.72	75.97	65.00
4	10.0	1132.20	1214.77	1029.63	75.88	73.97	62.43

Table 4 Variation in air flow rate, fuel consumption rate, gas flow rate at air outlet temperature of the heat exchanger

Air flow rate (m^3/h)	Fuel consumption rate (kg/h)	Gas flow rate (m^3/h)	Air outlet temperature (°C)		
			Maize cobs	*Prosopis juliflora*	Saw dust briquettes
200	4.0	10.06	146	154	139
	6.0	15.23	167	162	146
	8.0	20.40	174	174	149
	10.0	25.57	181	181	153
300	4.0	10.06	140	149	135
	6.0	15.23	149	160	141
	8.0	20.40	153	170	148
	10.0	25.57	156	180	151
400	4.0	10.06	136	142	131
	6.0	15.23	141	155	138
	8.0	20.40	143	166	142
	10.0	25.57	148	172	149
500	4.0	10.06	134	139	128
	6.0	15.23	139	151	132
	8.0	20.40	149	158	138
	10.0	25.57	151	163	144

Effect of Air Flow Rate, Fuel Consumption Rate on LMTD, Heat Transfer Area, and Heat Exchanger Effectiveness Using Different Biomass

The variation in LMTD (log mean temperature difference) of the heat exchanger with fuel consumption rate and air flow rate using different biomass is shown in Fig. 5. The LMTD of the heat exchanger was increased with increase in fuel consumption rate but decreased with increase in air flow rate. The effect of

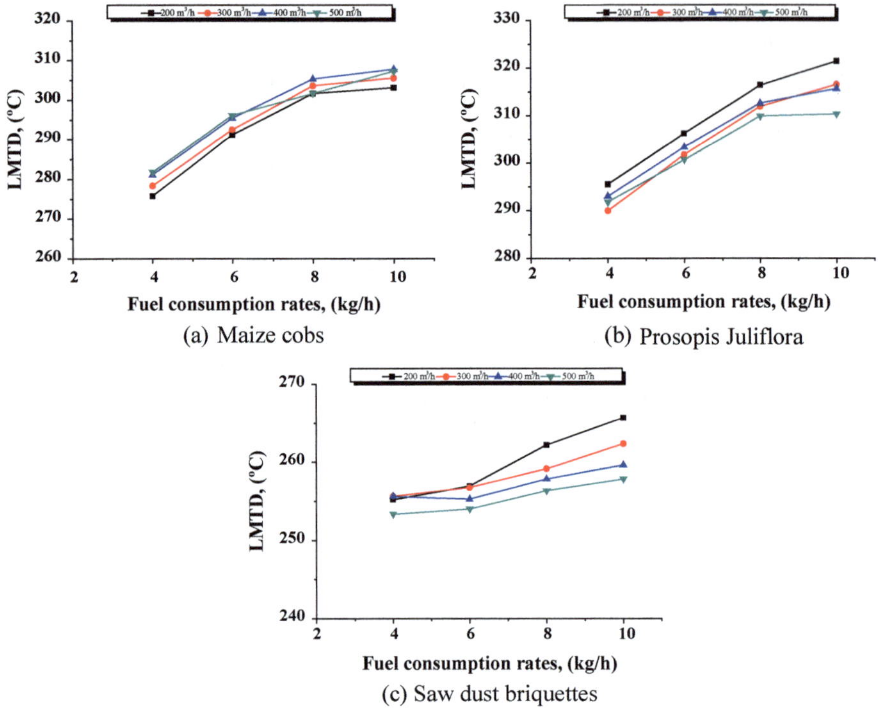

Fig. 5 Variation in LMTD of the heat exchanger with fuel consumption rate and air flow rate using different biomass

interaction of these two factors on LMTD was found statistically significant on 5% level of significance.

The variation in heat transfer area of the heat exchanger with fuel consumption rate and air flow rate using different biomass are shown in Fig. 6. The value of heat transfer area of the heat exchanger was decreased with increase in fuel consumption rate but increased with increase in air flow rate. The air flow rate and fuel consumption rate increased, the heat transfer area was found statistically significant on 5% level of significance.

The variation of heat exchanger effectiveness with fuel consumption rate and air flow rate using different biomass are shown graphically in Fig. 7. The heat exchanger effectiveness was varied in between 60.99–70.45; 64.87–74.88 and 68.38–74.73% at all the levels of fuel consumption rates, i.e., 4, 6, 8, and 10 kg/h and air flow rates, i.e., 200, 300, 400, and 500 m^3/h using maize cobs, *P. juliflora*,

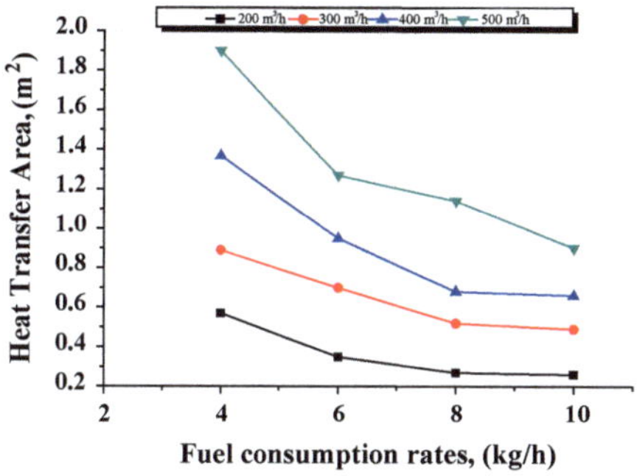

(a) Maize cobs

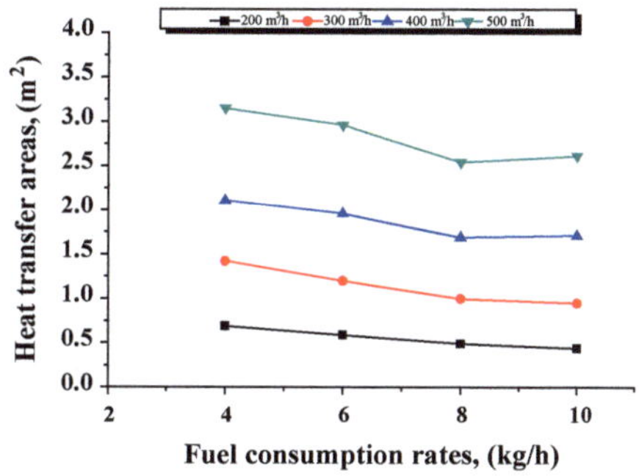

(b) Prosopis Juliflora

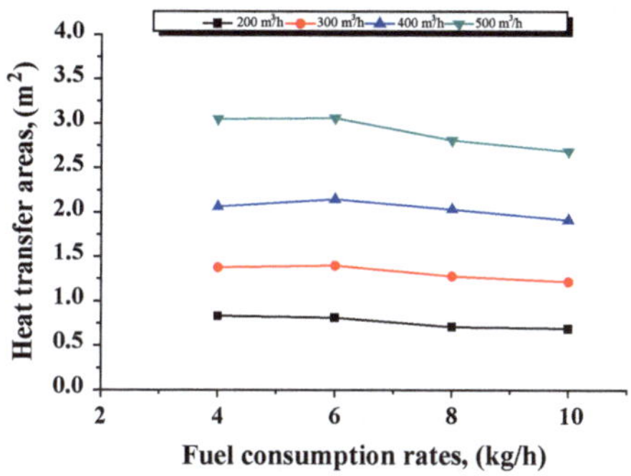

(c) Saw dust briquettes

Fig. 6 Variation in heat transfer area of the heat exchanger with fuel consumption rate and air flow rate using different biomass

and saw dust briquettes, respectively. The minimum heat exchanger effectiveness 60.99, 64.87, and 68.38% was obtained at 10 kg/h fuel consumption rate and 200 m^3/h air flow rate. The maximum heat exchanger effectiveness 70.45, 74.88, and 74.73% was obtained at 4 kg/h fuel consumption rate and 500 m^3/h air flow rate. The heat exchanger effectiveness increased with increase in air flow rate and decreased with decrease in fuel consumption rate.

Effect of Air Flow Rate and Fuel Consumption Rate on Dryer Inlet Hot Air Temperature Using Different Biomass

The variation in dryer inlet hot air temperature with fuel consumption rate and air flow rate using different biomass is shown in Fig. 8. The dryer inlet hot air temperatures were varied in between 66–89; 69–82 and 62–83 °C at all the levels of

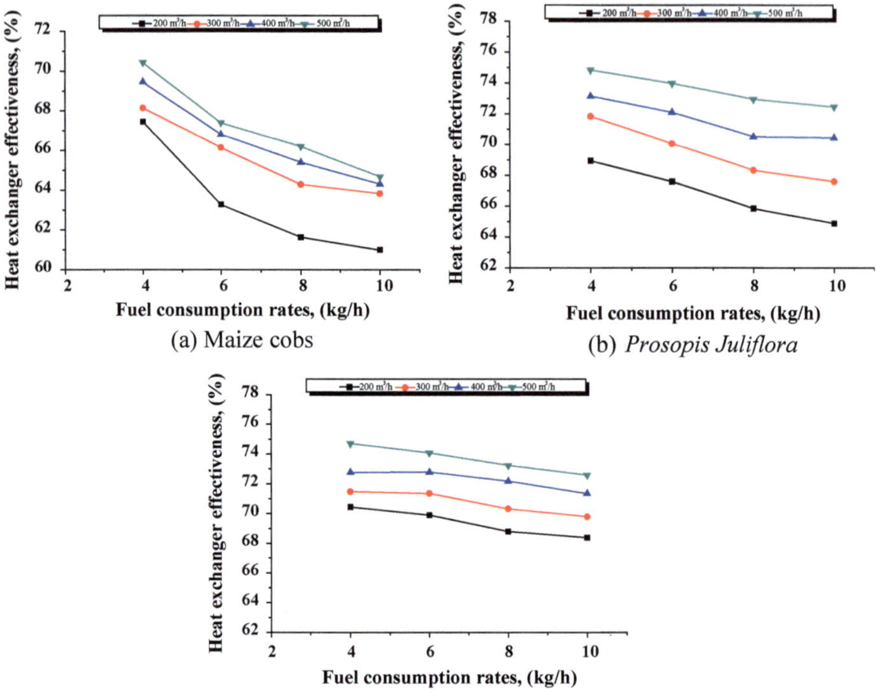

(a) Maize cobs (b) *Prosopis Juliflora* (c) Saw dust briquettes

Fig. 7 Variation in heat exchanger effectiveness with fuel consumption rate and air flow rate using different biomass

fuel consumption rates, i.e., 4, 6, 8, and 10 kg/h and air flow rates, i.e., 200, 300, 400, and 500 m^3/h using maize cobs, *P. juliflora*, and saw dust briquettes, respectively. The dryer inlet hot air temperatures were increased with increase in fuel consumption rates but decreased with increase in air flow rates. The air flow rate and fuel consumption rate were increased, the dryer inlet hot air temperature was found statistically significant on 5% level of significance.

Conclusion

The producer gas-based heat exchanger was designed and evaluated using different biomass for drying application. The physical properties and proximate analysis of the different biomass show good fuel value for the developed gasifier. Diameter of reactor, height of reactor, area required, and volume of reactor were computed for maize cobs, *P. juliflora*, and saw dust briquettes used in the updraft biomass gasifier as 0.40 m, 1.00 m, 0.1256 m^2, and 0.1256 m^3, respectively. The amount of carbon monoxide (CO), hydrogen (H_2) and methane (CH_4), and calorific value of producer

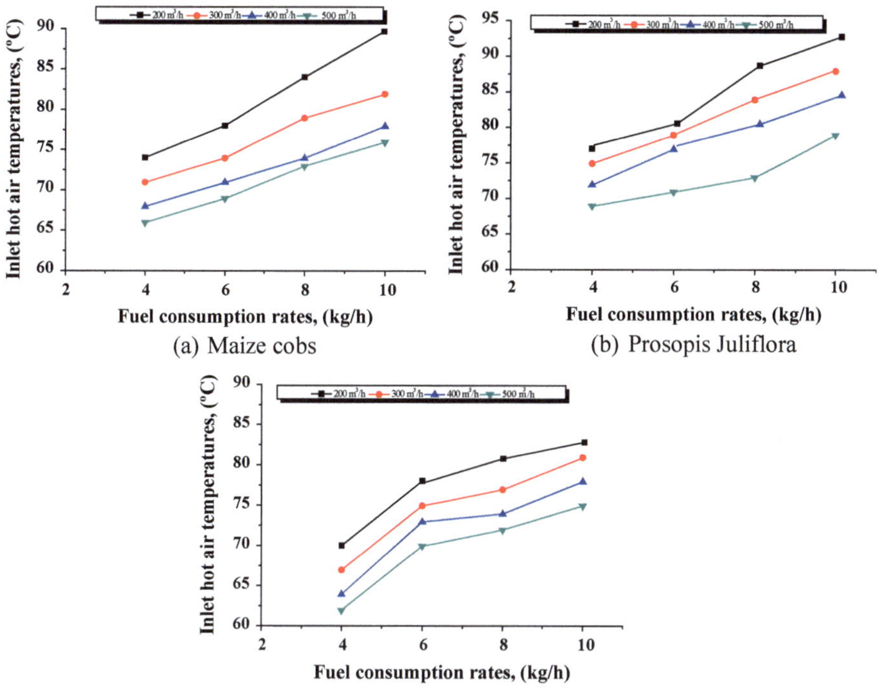

(a) Maize cobs (b) Prosopis Juliflora (c) Saw dust briquettes

Fig. 8 Variation in dryer inlet hot air temperatures with fuel consumption rate and air flow rate using different biomass

gas were decreased with increase in fuel consumption rate. The minimum gasifier efficiency of 75.88, 73.97, and 62.43% was found at the fuel consumption rate of 10 kg/h respectively using maize cobs, *P. juliflora*, and saw dust briquettes as fuel in the system. The minimum and maximum heat exchanger effectiveness 60.99, 64.87, and 68.38% and 70.45, 74.88, and 74.73% was obtained at 10 and 4 kg/h fuel consumption rate and 200 and 500 m^3/h air flow rate, respectively, using maize cobs, *P. juliflora*, and saw dust briquettes. The minimum hot air temperature of the dryer 66, 69, and 62 °C was measured at 4 kg/h fuel consumption rate and 500 m^3/h air flow rate, respectively, using maize cobs, *P. juliflora*, and saw dust briquettes. The maximum hot air temperature 89, 92, and 83 °C was measured at 10 kg/h fuel consumption rate and 200 m^3/h air flow rate using maize cobs, *P. juliflora*, and saw dust briquettes, respectively. From the above, it can be concluded that the producer gas base heat exchanger is good for thermal application.

References

Alattab KA, Zainal ZA (2006) Externally fired gas turbine for small scale power unit/hot air production unit for drying processes using biomass fuel. In: International conference on energy and environment, 28–29 Aug 2006. University Tenaga National, Kuala Lumpur, Malaysia

ASABE (2006) ANSI/ASAE S424.1—method of determining and expressing particle size of chopped forage materials by screening. ASABE Standards, St. Joseph, MI:—American Society of Agricultural and Biological Engineers, pp 619–621

ASTM (1983) Annual book of ASTM standards. American Society for Testing of Materials, Philadelphia, p 19103

Bergman RD (2005) Small scale lumber drying using wood gasification as heat source. Western Dry Kiln Association, pp 40–46

Kern DQ (1965) Process heat transfer, Int. edn. McGraw-Hill Book Company

Perry JH (1963) Chemical engineer's handbook, 4th edn. McGraw Hill book Company, New York

Rabou LPLM, Grift JM, Conradie RE, Fransen S, Verhoeff F (2007) Micro gas turbine operation with biomass producer gas. In: Contribution to the 15th european biomass conference, Berlin

Silva JND, Zanata FL, Scholz V, Galvarro SFS (2014) Biomass gasifier coupled to a gas combustion chamber and it's potential for thermal conditioning of poultry house and grain drying. In: International conference of agricultural engineering (AgEng), Zurich, 06–10.07.2014. www.eurageng.eu

Vasudevan P, Sharma S, Kumar A (2005) Liquid fuel from biomass: an overview. J Sci Ind Res 64:822–831

Vera D, Jurado F, Carpio J (2011) Study of a downdraft gasifier and externally fired gas turbine for olive industry wastes. Fuel Process Technol 92:1970–1979

Vyas DK (2003) Studies on biomass gasification for power generation. Unpublished M.Tech. thesis, College of Agricultural Engineering and Technology, Gujarat Agricultural University, Junagadh

Zainal ZA (1996) Performance and characteristics of a biomass gasifier. Ph.D. thesis, University of Wales, Cardiff, United Kingdom

Carbon Storage Potential in Dominant Trees of Koraput District of Odisha

Kakoli Banerjee, Gopal Raj Khemendu, Rakesh Paul and Abhijit Mitra

Abstract A survey was conducted during 2015 in major terrestrial trees of Koraput with the aim to evaluate the biomass and stored carbon in the species. The order of biomass and stored carbon varied as per the sequence *Mangifera indica* > *Pongamia glabra* > *Tamarindus indica* > *Eugenia jambolana* > *Shorea robusta* > *Artocarpus heterophyllus* > *Bombax malbaricum* > *Santalum album* > *Anacardium occidentale.* Correlation coefficient conducted between DBH and tree biomass and stored carbon indicates significantly positive interrelationship between the variables. However, in case of interrelationship between DBH and height, the same trend was observed except in species *Anacardium* sp., *Bombax* sp., and *Tamarindus* sp. The overall result signifies that greater biomass of trees serves as potential reservoir of carbon and may be used to off-set the carbon dioxide concentration at local level.

Keywords Koraput · Diameter at breast height (DBH) · Biomass
Stored carbon (C)

Introduction

Forest ecosystems can be said to be the net sources or sinks of CO_2, depending on dominant biological or physical factors, including: (1) state of the soil and vegetation (i.e., the system undisturbed, disturbed or recovering?); (2) management practices at the site level; (3) environmental conditions (e.g., climatic, edaphic, fire, pests, etc.); and (4) atmospheric deposition of pollutants and other compounds, some of which (e.g., CO_2 and nitrogen) can serve as nutrients (Kauppi et al. 1992;

K. Banerjee (✉) · G.R. Khemendu · R. Paul
Department of Biodiversity and Conservation of Natural Resources,
Central University of Orissa, Landiguda, Koraput 764020, Odisha, India
e-mail: banerjee.kakoli@yahoo.com

A. Mitra
Department of Marine Science, University of Calcutta, 35 B.C. Road,
Kolkata 700019, West Bengal, India

V.P. Singh et al. (eds.), *Energy and Environment*, Water Science
and Technology Library 80, https://doi.org/10.1007/978-981-10-5798-4_22

Cropper and Gholz 1993). It has been reported by Winjum et al. (1992) that there are three broad classes of forest management actions that could influence carbon conservation and sequestration in forest ecosystems: (1) decreasing deforestation and forest degradation; (2) establishing additional areas of forest; and (3) implementation of practices which stimulate CO_2 fixation by existing forest or agro-forest systems. About 80% of the world's potential for increasing carbon storage in forests (estimated at 60–87 Pg carbon from now up to 2050) lies in developing countries (Brown 1995). The forest woodland contains more carbon than all other categories collectively, but the proportion of total carbon in this class has progressively declined throughout the century from 73% in 1880 to 63% in 1980 (Richards and Flint 1993). While as, in India, a national level estimate of carbon storage conducted by Kishwan et al. (2009) estimated the role of India's forests from 1995 to 2005 toward carbon (*C*) sink using secondary data of growing stock from different sources and reported that from 1995 to 2005, carbon in biomass of Indian forests have increased from 2692.474 to 2865.739 mt registering an annual increment of 173.265 mt of carbon during a decade. Sathaye and Ravindranath reported in 2001 on Climate Change that about 36.9 million hectare degraded forestland with carbon mitigation potential of 74.75 t carbon/ha is available for regeneration. Moreover, India with carbon abatement cost in the forestry sector can be the basis of attracting global environment facility (GEF) funded projects.

The gases with special optical properties that are responsible for climate warming include carbon dioxide (CO_2), water vapors (H_2O), methane (CH_4), nitrous oxide (N_2O), nitrogen oxides (NO_x), stratospheric ozone (O_3), carbon monoxide (CO), and chlorofluorocarbons (CFC's). Among all these greenhouse gases, CO_2 plays a lead role as it contributes to 50% of the total greenhouse effect (Bhardwaj and Panwar 2003). Though fundamental to life on earth, concentration of atmospheric CO_2 has increased from a preindustrial level of 280 to 390 ppm and is increasing at a rate of 1.5 $\mu L\ year^{-1}$ giving rise to an alarming situation (Lee and Dobson 1996). Anthropogenic activities, including combustion of fossil fuels and land use change, contribute carbon dioxide (CO_2) emissions to the global carbon (*C*) cycle. Current annual CO_2 emissions are estimated at 7.5 ± 1.5 petagrams ($Pg = g \times 10^{15}$) as *C* (Sundquist 1993). The atmospheric CO_2 reservoir is increasing by 3.4 ± 0.2 Pg annually (Tans et al. 1990). The world's oceans are believed to be a net sink of 2.0 ± 0.8 PgC annually, leaving the terrestrial biosphere as a sink up to 4 PgC annually (Tans et al. 1990; Sundquist 1993). Human activities (like expansion of mining areas, clearing of the forest areas for human settlement, industrialization, and urbanization) are also responsible for making changes in carbon stocks in these pools by changing the land use pattern of any area (Negi and Chauhan 2002). They have caused a significant release of CO_2 to the atmosphere from the terrestrial biota and soils, soil being a major source of atmospheric CO_2.

Terrestrial ecosystems of the northern hemisphere have been identified as potential carbon sink (Tans et al. 1990; Kauppi et al. 1992; Sedjo 1992), although some tropical forests are also accumulating significant amounts of carbon (Brown et al. 1992). However, the precise role of the terrestrial biosphere in the global carbon cycle still remains uncertain. It is to be noted here that the challenges of climate change can be effectively overcome by the storage of carbon in terrestrial carbon sinks, viz., plants, plant products, and soils for longer periods of time. Forestry can play a major role towards increasing the global carbon sequestration if the world's forest could be managed properly with due importance to afforestation and reforestation and carbon management in existing forests (Bala et al. 2003; McKinley et al. 2011). Forest transition also contributes to carbon sequestration besides conserving biodiversity and improving local and regional environment (Xu et al. 2007). On this background, this first-order analysis aims to estimate the biomass and carbon storage potential in dominant trees of Koraput district of Odisha which is noted for its deciduous variety of trees at a location of approximately 853.44 m from the mean sea level (MSL).

Materials and Methods

Study Area

Koraput has a population of 13,05,492 (2008) spread in an area of 8,807 km^2, and is situated in the southern part of Odisha State (18° 13′ and 19° 10′ North latitude and 82° 5′ and 83° 13′ East longitude) with a total geographical area of 8,807 km^2 accounting 5.38% of Odisha state. Around 21% (1879.53 km^2) of the total area is covered with forest (Sahu et al. 2003; DES 2007). Physiographically it is contiguous to the main land of Eastern Ghat high land zone and South-Eastern Ghat zone. It is bounded in the east by Rayagada district (a portion) of Orissa and Srikakulam district of Andhra Pradesh, Bastar district of Chhattisgarh in the West, Nabarangpur district of Orissa in the north and Malkangiri district of Orissa and Visakhapatnam district of Andhra Pradesh in the South. The general topography of the area is of broken mountains intercepted by large riverbeds and watercourses. The altitude varies from 500 m near western side to 1600 m on the eastern side with mountain peaks and ridges. Deomali (1672 m) is the highest mountainous peak of Orissa found in this district. Sandy and clay type soil predominate the entire district. The climate of the major portion of the district is influenced by its varied elevation. The minimum and maximum temperatures are 13 and 42 °C in the month of December and May, respectively. Humidity is generally high especially in the monsoon and post-monsoon months. It receives about 1500 mm rainfall annually (Dash 1994). Major portion of the annual rainfall during southwest monsoon occurs between July and September. The forest vegetation of Koraput is broadly divided into three categories viz. semi evergreen, moist deciduous, and bushy type (Das and Misra 2000). Five stations were selected at random and five quadrants each of 25 m × 25 m were selected in each station to study the biomass and carbon of the tree species (Table 1 and Fig. 1).

Table 1 Study site with description

S. No.	Station name	Coordinates	Description
1	Tarlaguda	18° 44′ 38.2″N 82° 54′ 54.7″E	It is a social forest near Damanjodi town plant are protected by the local villagers. It is partially prone to pollution as it near the NALCO industry
2	Marchimaal	18° 44′ 07.3″N 82° 54′ 36.2″E	The forest almost 5 km away from Damanjodi town and also protected by the local villager. Two stone mining stations are present near the site
3	Doliambo	18° 39′ 24.4″N 82° 52′ 45.5″E	The station is away from Semiliguda town, also near to a mining industry. A sal plantation forest is present beside the patch of forest
4	Amtiguda	18° 47′ 02″N 82° 44′ 57″E	Present near the Koraput town, and the plants are protected by the local people since the past 60–70 years due to their religious belief and social benefits
5	Sukriguda	18° 48′ 51.2″N 82° 40′ 07.4″E	A natural forest nearly 10 km away from the Koraput town and away from the pollution of the road vehicles. Present adjacent to village Sukriguda, present opposite to a Eucalyptus plantation

Sampling and Analysis

We selected 09 dominant tree species in the area based on the relative density and abundance of the trees. The trees were mostly from social forestry which is wild and >20 cm diameter was taken into consideration. The age of the trees were >30 years. The study was undertaken in October to December 2015 (post-monsoon season). Above ground biomass (AGB) estimation: The biomass of above ground structures (stems, branches, and leaves) was estimated as per the standard procedure (Mitra et al. 2011). Above ground carbon (AGC) estimation: The fresh samples of stem, branch, and leaf were collected for each species and oven dried at 70 °C. Direct estimation of percent carbon for each species were carried out by *Vario MACRO elementar* make CHN analyzer, after grinding and random mixing the oven dried stems, branches and leaves separately (Mitra et al. 2011).

Results and Discussion

A total of 09 species has been documented from the study site (Table 2). The relative densities of the trees at the five selected stations are given in Fig. 2. From the abundance chart, it is clearly seen that *Mangifera* sp. holds the major share followed by *Pongamia* sp., *Eugenia* sp. and *Shorea* sp.

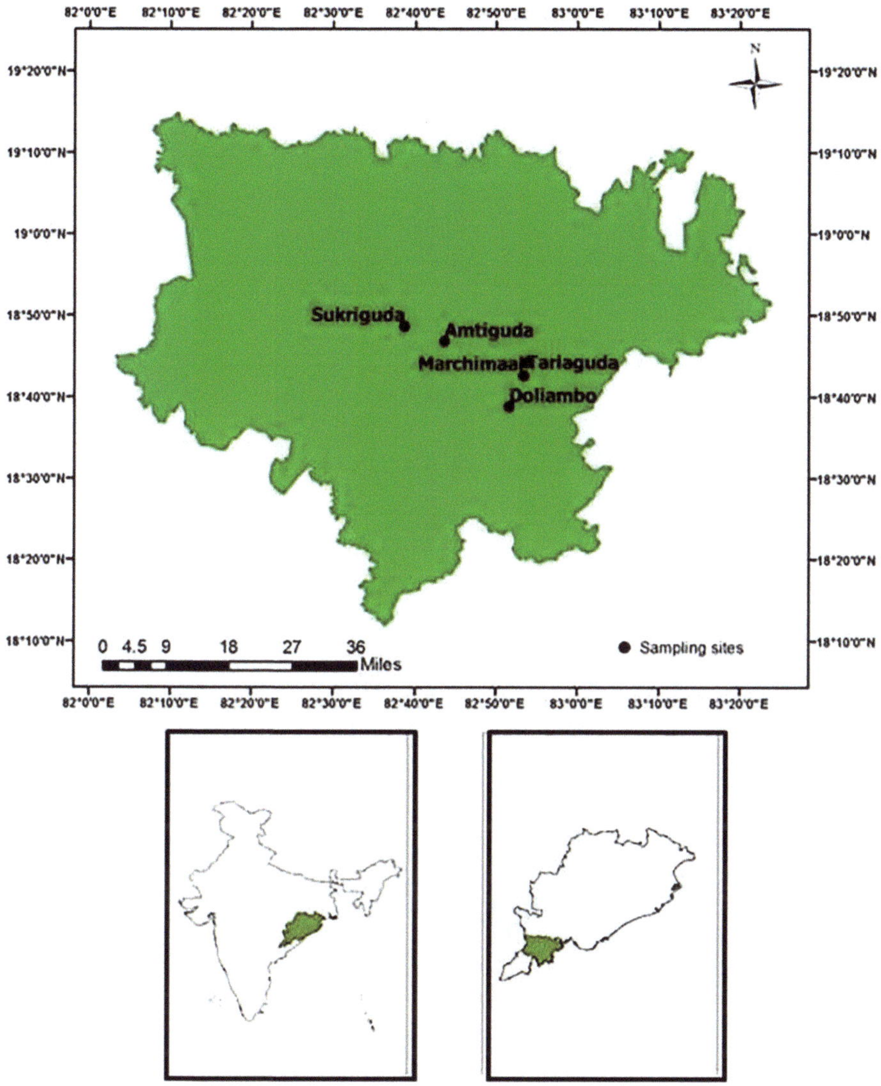

Fig. 1 Map showing the location of sampling stations

Above Ground Biomass (AGB)

Biomass data is a basic requirement for the estimation of carbon density and storage and can be acquired in different ways but field measured data is the most basic, direct, and authentic (Wang et al. 1999; Fang et al. 2001). At the individual plant level, photosynthesis is dependent on ambient CO_2 concentration, light, temperature, and other factors (Mooney et al. 1991; Cropper and Gholz 1993). Strain and

Table 2 Taxonomic identification of species with their AGB and AGC values in the study area

S. No.	Systematic position	Species		Stations				
				1	2	3	4	5
1	*Anacardium occidentale* Kingdom-plantae order-Myrtales family-Anacardiaceae genus-*Anacardium species-occidentale*		AGB (kg/ha)	80.64 ± 1.15	–	110.24 ± 0.83	–	554.56 ± 0.64
			AGC (47.5%)	38.30 ± 0.54	–	52.36 ± 0.39	–	263.42 ± 0.30
2	*Artocarpus heterophyllus* Kingdom-planate order-Rosale family-Moraceae genus-*Artocarpus* species-*heterophyllus*		AGB (kg/ha)	2577.44 ± 4.08	6717.44 ± 9.78	10495.84 ± 12.28	3439.04 ± 9.63	4614.72 ± 4.59
			AGC (48.3%)	1244.9 ± 1.97	3244.52 ± 4.72	5069.49 ± 5.93	1661.06 ± 4.65	2228.91 ± 2.22
3	*Mangifera indica* Kingdom-plantae order-Sapindles family-Anacardiaceae genus-*Mangifera* species-*indica*		AGB/ha)	22289.44 ± 4.52	181003.70 ± 5.68	1131.20 ± 3.86	16314.88 ± 6.30	6798.08 ± 3.76
			AGC (47.8%)	10654.35 ± 2.16	86519.76 ± 2.71	540.71 ± 1.84	7798.51 ± 3.01	3249.48 ± 1.80

(continued)

Table 2 (continued)

S. No.	Systematic position	Species		Stations				
4	*Bombax malabaricum* Kingdom-plantae order-Malvales family-Bombacaceae genus-*Bombax* species-*malabaricum*		AGB (kg/ha)	5127.52 ± 3.29	3612 ± 3.05	–	5088.16 ± 2.88	13736.80 ± 5.46
			AGC (47.2%)	2420.19 ± 1.55	1704.86 ± 1.44	–	2401.61 ± 1.36	6483.77 ± 2.58
5	*Pongamia glabra* Kingdom-plantae order-Fables family-Fabaceae genus-*Pongamia* species-*glabra*		AGB (kg/ha)	–	11592.48 ± 4.42	–	53872.64 ± 4.63	16211.84 ± 4.77
			AGC (47.9%)	–	5552.8 ± 2.12	–	25804.99 ± 2.22	7765.47 ± 2.28
6	*Tamarindus indica* Kingdom-plantae order-Fables family-Fabaceae genus-*Tamarindus* species-*indica*		AGB (kg/ha)	9056.8 ± 12.48	14989.12 ± 8.51	–	25102.24 ± 7.03	9159.52 ± 5.75
			AGC (48.4%)	4383.49 ± 6.04	7254.73 ± 4.12	–	12149.48 ± 3.41	4433.21 ± 2.78

(continued)

Table 2 (continued)

S. No.	Systematic position	Species		Stations				
7	*Eugenia jambolana* Kingdom-plantae order-Myrtales Family-Myrtaceae genus-*Eugenia* species-*jambolana*		AGB (kg/ha)	17765.92 ± 3.68	7675.36 ± 5.22	2998.56 ± 10.57	23735.36 ± 6.84	4968.96 ± 4.82
			AGC (49.2%)	8740.83 ± 1.81	3776.28 ± 2.57	1475.29 ± 5.20	11677.8 ± 3.36	2444.73 ± 2.37
8	*Shorea robusta* Kingdom-plantae order-Malvales family-Dipterocarpaceae genus-*Shorea* species-*robusta*		AGB (kg/ha)	–	–	563.84 ± 6.51	23809.44 ± 11.03	28790.1 ± 8.26
			AGC (49.5%)	–	–	279.1 ± 3.22	11785.67 ± 5.46	14251.1 ± 4.09
9	*Santalum album* Kingdom-plantae order-*antalales* family-Santalaceae genus-*Santalum* species-*album*		AGB (kg/ha)	–	–	522.08 ± 5.80	281.92 ± 7.01	282.56 ± 4.59
			AGC (48.6%)	–	–	253.73 ± 2.82	137.01 ± 3.40	137.32 ± 2.23

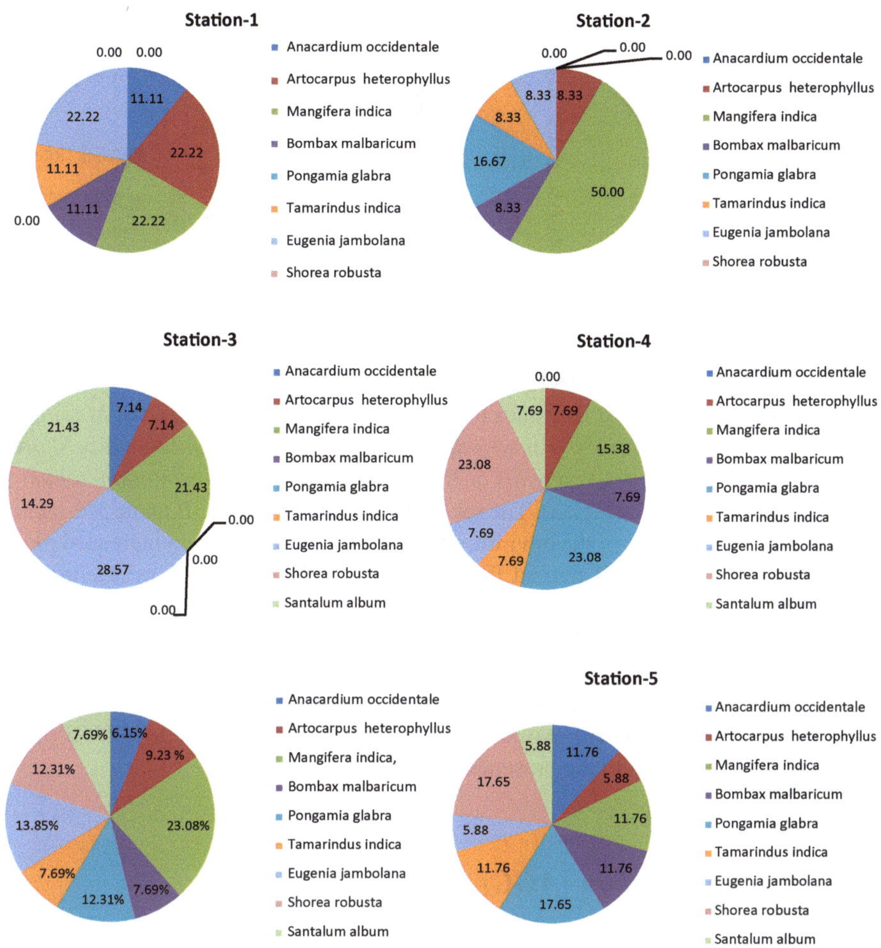

Fig. 2 Relative density of the trees at the selected stations and abundance of tree species considering all the stations

Thomas (1992) reviewed the literature on plant response to elevated CO_2 and concluded: (1) if other resources are present at required levels, CO_2 enrichment will increase photosynthesis and plant growth; (2) plants limited by resource deficiencies (e.g., nitrogen or phosphorus) will respond slightly or not at all to CO_2 enrichment; (3) CO_2 reduces transpiration and improves plant water status, due to increasing photosynthesis and decreased water loss; and (4) CO_2 and global warming may affect species differentially and will result in ecosystem flora and fauna change. The potential of individual trees to act as a carbon sink may be highly dependent on response to soil nutrition and environmental stress rather than to atmospheric CO_2 concentration (Norby et al. 1992). Forest ecosystem carbon

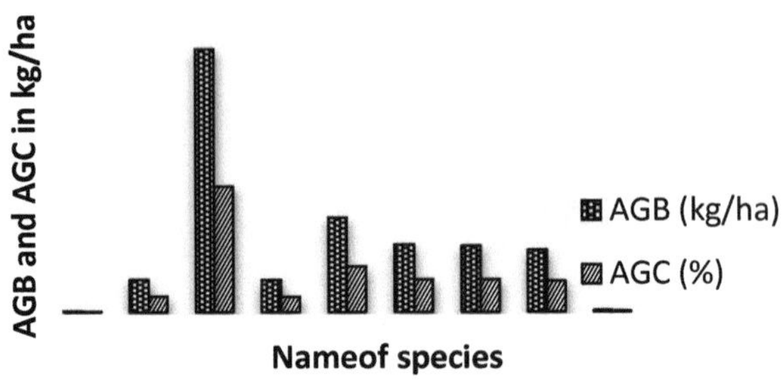

Fig. 3 Species-wise variation in AGB and AGC values in the study area

balances appear to be sensitive to annual differences in climate and possibly CO_2 enrichment (Mooney et al. 1991; Cropper and Gholz 1993).

This study shows that AGB values in all the stations follow the trend of *Mangifera indica* > *Pongamia glabra* > *Tamarindus indica* > *Eugenia jambolana* > *Shorea robusta* > *Artocarpus heterophyllus* > *Bombax malbaricum* > *Santalum album* > *Anacardium occidentale* (Fig. 3). The relatively more value of AGB in the *Mangifera* sp. may be the suitability of the environment in course of time and an increased pH of the ambient soil. The AGB ranged from 80.64 ± 1.15 kg/ha (in *A. occidentale*) at station 1 to 181,003.70 ± 5.68 kg/ha (in *M. indica*) at station 2 considering minimum and maximum values of AGB respectively during the study period. The total biomass in each station varied as Station 2 (225590.1 kg/ha) > Station 4 (151643.68 kg/ha) > Station 5 (85117.14 kg/ha) > Station 1 (56897.76 kg/ha) > Station 3 (15821.76 kg/ha). The highly acidic soil may be one of the reasons behind the low AGB of the species at stations 3 due to the presence of the mining industry.

Above Ground Carbon (AGC)

In terrestrial system, carbon is retained in live biomass, decomposing organic matter and soil that play an important role in the global carbon cycle. Carbon is exchanged between these systems and the atmosphere through photosynthesis, respiration, decomposition, and combustion. Adoption of carbon sequestration measures in the forests and soil can considerably reduce the rise in atmospheric CO_2 level (Rai and Sharma 2003). In order to sustain the amount of carbon in the soil, the identified ecological factors should be enhanced through the application of good forest and land management practices, such as creation of vegetal buffer zones around farmlands, zero-tillage practice, mulching, retaining of forest slash and crop residues, fertilizer application, elongation of fallow periods, crop rotation, and tree planting initiatives in degraded areas among others. Through these healthy practices, forest vegetation can be maintained; thereby increasing the carbon stock of forest soil by reducing direct loss to the atmosphere (Offiong and Iwara 2012). A study carried out on carbon sequestration and carbon sink by Gera et al. 2002 in community projected forests of Sambalpur forest division, Orissa, India showed that

1.53–3.01 tons of carbon is being sequestered per ha per year, with only protection, which can be enhanced through proper implementation of the management prescriptions. The study also suggested that Joint Forest Management in India could be effectively utilized for carbon sequestration so as to mitigate climate change.

In this study, the stored carbon in the tree is found to increase with the growth of species (documented in the present study in terms of biomass). The AGC values ranged from 38.30 ± 0.54 kg/ha (in *A. occidentale*) to 86,519.76 ± 2.71 kg/ha (in *M. indica*) during the study period considering the minimum and maximum values of AGC. The percentage of carbon in the selected species ranged from 47.2% (in *B. malbaricum*) to 49.5% (in *S. robusta*) during the post-monsoon season (Table 2). The station wise variation in AGC is shown in Fig. 4. Considering the overall accumulation pattern of carbon in each station, we can conclude that Station 2

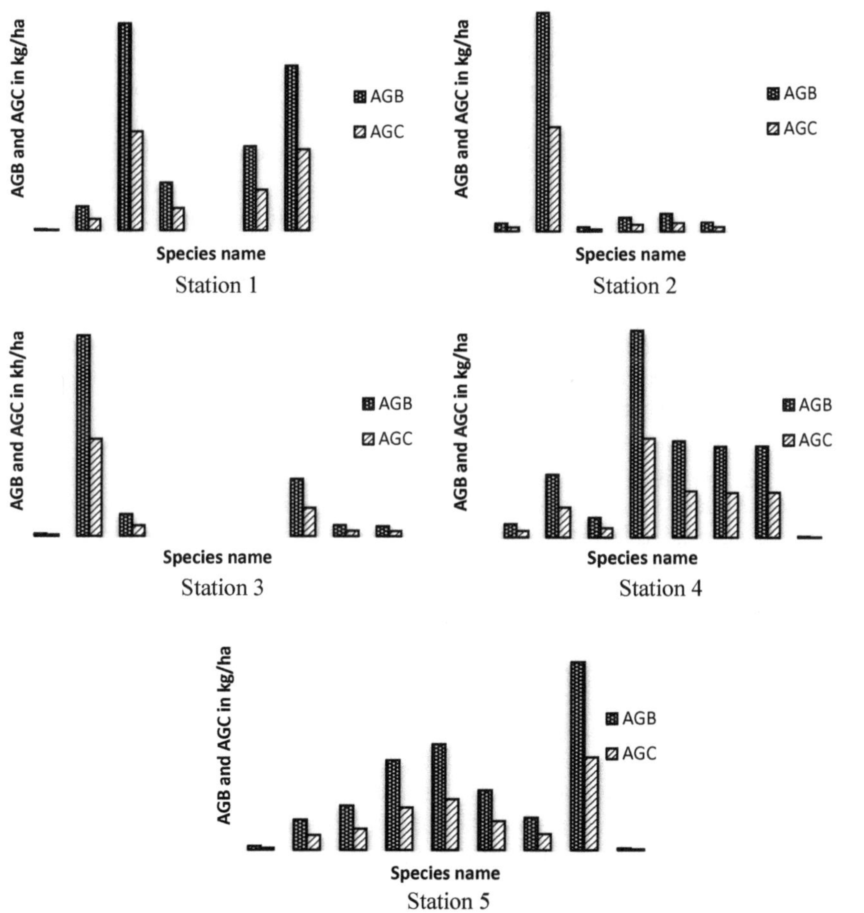

Fig. 4 Variation in AGB and AGC values in the selected species at the sampling stations

Table 3 Inter-relationship (*r*-value) between DBH, height, biomass and carbon of selected tree species

Species	DBH × height	DBH × biomass	Biomass × carbon
Anacardium occidentale	−0.63, p <0.01	0.90, p <0.01	0.90, p <0.01
Artocarpus heterophyllus	0.52, p <0.05	0.94, p <0.01	0.94, p <0.01
Mangifera indica	0.90, p <0.01	0.92, p <0.01	0.92, p <0.01
Bombax malabaricum	−0.17, IS*	0.83, p <0.01	0.83, p <0.83
Pongamia glabra	0.83, p <0.01	0.95, p <0.01	0.94, p <0.01
Tamarindus indica	0.13, IS*	0.86, p <0.01	0.86, p <0.01
Eugenia jambolana	0.91, p <0.01	0.95, p <0.01	0.95, p <0.01
Shorea robusta	0.95, p <0.01	0.96, p <0.01	0.96, p <0.01
Santalum album	0.63, p <0.01	0.96, p <0.01	0.95, p <0.01

*IS means insignificant

(108052.95 kg/ha) > Station 4 (73416.13 kg/ha) > Station 5 (41257.41 kg/ha) > Station 1 (27482.06 kg/ha) > Station 3 (7670.68 kg/ha) respectively.

We deviated from the standard norm of expressing the carbon content in tons ha^{-1}, because this value is the product of number of trees of a particular species and their respective volumes, which depend on human choice during plantation process and growth (in terms of biomass), where management is also a vital component (Zaman et al. 2014). Correlation coefficient conducted between DBH and tree biomass and stored carbon indicates significantly positive interrelationship between the two variables (Table 3). However, in case of interrelationship between DBH and height the same trend was observed except in species *Anacardium* sp., *Bombax* sp., and *Tamarindus* sp. The overall result signifies that greater biomass of trees serves as potential reservoir of carbon and may be used to mitigate the carbon dioxide concentration at local level.

A comparison of carbon sequestration rates and total carbon stock pile in degraded and nondegraded sites of Oak and Pine forest of Kumaun Central Himalaya was made by Jina et al. (2008). The study confirms that the sequestration of CO_2 in nondegraded forests is significantly greater than the degraded forests. The study further suggests that community forests should be encouraged because of their significance of becoming the sink for increased CO_2 worldwide. This is also at par with our study which is the social forests and purely conserved by the local people. Plant species which are less vulnerable to climate change have a greater potential of carbon sequestration. This view is supported by Negi and Chauhan (2002), who studied the greenhouse gas mitigation potential by Sal (*S. robusta* Gaertn F.) forests in Doon valley, Uttaranchal, India. Sal, a less vulnerable species to climate change provides a positive response to climate change shown by its increased productivity, and thus has the capability to sequester greater amounts of carbon. Davey et al. (2006) conducted a study on popular trees and came to the conclusion that these trees are well suited to elevated levels of CO_2 and can be grown for long term storage of carbon in wood.

Koraput although is a small city sustains two major industries of Hindustan Aeronautics Limited (HAL) and National Aluminium Company (NALCO) within its vicinity. Though the area is less populated for now but the two major industries can be a probable cause in increase of carbon dioxide levels in the atmosphere. Such studies at the local level aim to prepare a stand and state level carbon budget for the state and to describe the temporal carbon dynamics of these forests. This fact is also supported by Pande (2003) who did a similar survey in the natural forests of Madhya Pradesh, India. The study reveals that open canopy forests have a great potential for sequestrating more and more carbon.

Conclusions

In today's world, there is an utmost need of developing a data bank on potential forests as a sink of carbon. There is no doubt that awareness has already developed on adverse impact of greenhouse gas emission and the consequent climate change. At the dawn of third millennium, greenhouse gases are widely accepted by international scientific community as one of the potential threats to the existence of human kind coupled with extinction of other flora and fauna. Sequestration and conservation of carbon in forest and agro-ecosystems can be achieved at relatively low initial cost (NAS 1991; Dixon et al. 1993). The implementation of such practices can lead to a positive rate of financial return as far as the concept of carbon credit is concerned.

Acknowledgements The authors duly acknowledge NRSC Hyderabad for their financial assistance for carrying out the research work. The authors also acknowledge the constant support of HOD and Departmental staff for their constant support.

References

Bala N, Singh G, Kumar P, Sinha AK (2003) Role of forests in carbon sequestration. Indian Forester 129:799–805

Bhardwaj SD, Panwar P (2003) Global warming and climate change-effect and strategies for its mitigation. Indian Forester 129:741–748

Brown S (1995) Management of forests for mitigation of greenhouse gas emissions. In: Houghton JT, Jenkins GJ, Ephraums JJ (eds) Climatic change, impacts, adaptations and mitigation of climate change: scientific-technical analyses. Cambridge University Press, Cambridge pp 773–797

Brown S, Lugo AE, Iverson LR (1992) Processes and lands for sequestering carbon in the tropical forest landscape. Water Air Soil Pollut 64:139–155

Cropper WP, Gholz HL (1993) Constructing a seasonal carbon balance for a forest ecosystem. Clim Res 3:7–12

Dash SS (1994) Ethnobotanical study of Narayanapatna area of Koraput district, Orissa. M. Phil. Dissertation, Berhampur University: Berhampur, Orissa

Das PK, Misra MK (2000) Vegetation and floristic studies on Koraput district of Orissa. Higher Plants of Indian Sub-continent, Bishen Singh Mahendra Pal Singh, Dehradun, India, ix, pp 115–130

Davey PA, Olcer H, Zakhleniuk O, Bernacchi CJ, Calfapietra C et al. (2006) Can fast-growing plantation trees escape biochemical down-regulation of photosynthesis when grown throughout their complete production cycle in the open air under elevated carbon dioxide?. Plant, Cell Environ 1–9

Directorate of Economics and Statistics (DES) (2007) District Statistical Handbook, Koraput, Bhubaneswar: Directorate of Economics and Statistics, Government of Orissa

Dixon RK, Andrasko KJ, Sussman FG, Lavinson MA, Trexler MC, Vinson TS (1993) Forest sector carbon offset projects: near-term opportunities to mitigate greenhouse gas emissions. Water, Air, and Soil Pollution 70:19–37

Fang JY, Chen AP, Peng CH, Zhao SQ, Ci LJ (2001) Changes in forest biomass carbon storage in China between 1949 and 1998. Science 292:2320–2322

Gera M, Bisht NS, Gera N (2002) Carbon sequestration through community based forest management—a case study from Sambalpur Forest Division, Orissa. Indian Forester 129:735–740

Jina BS, Sah P, Bhatt MD, Rawat, YS (2008) Estimating carbon sequestration rates and total carbon stockpile in degraded and non-degraded sites of oak and pine forest of Kumaun Central Himalaya. Ecoprint 15:75–81

Kauppi PE, Mielikainen K, Kuusela K (1992) Biomass and carbon budget of European forest, 1971 to 1990. Science 256:70–74

Kishwan J, Pandey R, Dhadwal VK (2009) India's forest and tree cover: contribution as a carbon sink. Indian Counc For Res Educ Bull 23:130p

Lee JJ, Dobson R (1996). Potential carbon sequestration by afforestation of pasture in the South-Central United States. Agron J 88:381–386

Mitra A, Sengupta K, Banerjee K (2011) Standing biomass and carbon storage of above-ground structures in dominant mangrove trees in the Sundarbans. Ecol Manage 261(7):1325–1335

McKinley DC, Ryan MG, Birdsey RA, Giardina CP, Harmon ME et al. (2011) A synthesis of current knowledge on forests and carbon storage in the United States. Ecol Adaptations 21:1902–1924

Mooney HA, Drake BG, Luxmoore RJ, Oechel WC, Pitelka LF (1991) How will terrestrial ecosystems interact with the changing CO_2 concentration of the atmosphere and anticipated climate change? BioScience 41:96–104

NAS (National Academy of Sciences) (1991) Policy implications of greenhouse warming. National Academy Press, Washington, DC

Negi JDS, Chauhan PS (2002) Greenhouse gases mitigation potential by Sal forest in Doon Valley. Indian Forester 128:771–778

Norby RJ, Gunderson CA, Wullschleger SD, O'Neill EG, McCracken MK (1992) Productivity and compensatory responses of yellow-poplar trees in elevated CO_2. Nature 357:322–324

Offiong RA, Iwara AI (2012) Quantifying the stock of soil organic carbon using multiple regression model in a fallow vegetation, Southern Nigeria. Ethiop J Environ Stud Manage 5:166–172

Pande PK (2003) Land carbon budget and sequestration potential of the natural forests of Madhya Pradesh. Indian Forester 129:905–917

Rai SC, Sharma P (2003) Carbon sequestration with forestry and land use/cover change: an overview. Indian Forester 129:776–786

Richards JF, Flint EP (1993) Historic land use and carbon estimates for South and South East Asia: 1880–1980. ORNL/CDIAC-61, NDP-46, Oak Ridge National Laboratory, Oak Ridge, Tennessee 404

Sahu SS, Gunasekharan K, Vanamail P, Jambulingam P (2003) Persistent foci of falciparum malaria among tribes over two decades in Koraput district of Odisha state, India. Malaria J 12:72. doi:10.1186/1475-2875-12-72

Sedjo RA (1992) Temperate forest ecosystems in the global carbon cycle. Arnbio 21:274–277

Strain BR, Thomas RB (1992) Field measurements of CO_2 enhancement and climate change in natural ecosystems. Water Air Soil Pollut 64:45–60

Sundquist ET (1993) The global carbon dioxide budget. Science 259:934–941

Tans PP, Fung IF, Takahashi T (1990) Observational constraints on the global atmospheric CO_2 budget. Science 247:1431–1438

Wang SQ, Zhou CH, Luo CW (1999) Studying carbon storage spatial distribution of terrestrial natural vegetation in China. Prog Geogr 18:238–244

Winjum JK, Dixon RK, Schroeder PE (1992) Estimating the global potential of forest and agroforest management practices to sequester carbon. Water Air Soil Pollut 64:213–228

Xu J, Yang Y, Fox J, Yang X (2007) Forest transition, its causes and environmental consequences: empirical evidence from Yunnan of Southwest China. Trop Ecol 48:137–150

Zaman S, Pramanick P, Pal N, Biswas P, Banerjee R, Biswas S, Bera N, Mitra A (2014) Stored carbon in dominant trees of urban area of Kolkata, India. Ideas Ideol Int e J 2(7):1–7

Study of Geothermal Energy Potential with Geothermal Doublet: A Case Study for Puga Valley Ladakh

Shibani K. Jha and Harish Puppala

Abstract The growing demands for renewable energy have made geothermal energy a popular option in recent past. The efficiency of such a system depends on the retention and transport capacity of heat. Hence, model studies under different reservoir conditions are essential. This study is concerned about a geothermal doublet to produce hot groundwater, extract heat, and reinject the cooled down water into the subsurface. Studies are performed by computational tool, COMSOL Multiphysics, to estimate the potential of geothermal reservoirs. The study shows the effect of geothermal doublet on coupled heat transport and groundwater flow. Puga Valley in Ladakh district, 1,600 km from New Delhi and at an altitude of about 4,400 m is considered for the study. The transient temperature distribution in the reservoir is discussed. The effect of natural groundwater flow on the temperature distribution and the influence of production and injection wells, the geothermal doublet, are also discussed.

Keywords Geothermal reservoir · Heat transport · Groundwater flow
Geothermal doublet · Injection well · Extraction well

Introduction

Geothermal energy provides an option for renewable energy for base load electricity in alleviating the world's energy and climate predicament (Ekneligoda and Min 2014). Geothermal potential, as regular steam and high temperature water, has been exploited for decades to produce power, and for space warming as well as for industrial purposes. The geothermal electrical plants were installed in the world to

S.K. Jha (✉) · H. Puppala
Department of Civil Engineering, Birla Institute of Technology and Science Pilani, Pilani 333031, India
e-mail: shibani@pilani.bits-pilani.ac.in

H. Puppala
e-mail: harishpuppala.ce@gmail.com

V.P. Singh et al. (eds.), *Energy and Environment*, Water Science and Technology Library 80, https://doi.org/10.1007/978-981-10-5798-4_23

produce 7974 MW till twentieth century. In the year 2000, with that installed capacity, an electrical energy of 49.3 billion kWh/year was generated, which was nearly 0.3% of the world aggregate electrical energy (Barbier 2002). Prior to 1864, geothermal energy resource in India was nearly unexplored. Beginning with an identification of 99 springs in India (Schlagintweit 1864), it was followed by the monumental work of exploring 340 springs (Rao1997). A major, systematic, multidisciplinary, and multi-Institutional programme (including drilling) was then mounted in India to explore the potential sites during 1972–74, covering the Puga–Chumatang field (Ladakh), India. From the existing literatures and the research works carried by geological survey of India (GSI), it is identified that the Puga geothermal reservoir is rich in hidden energy in the earth's crust which can feed the energy need of India at large scale.

The conventional way of extracting energy from the earth involves the process of injecting cold water through an injection well and extracting hot water from a production well (Schulte et al. 2010). This system of injection and extraction well is known as geothermal doublet. The distribution and frequency of this doublet in a reservoir influence the energy extraction potential. Since 1973, there has been developing interest for geothermal projects to utilize boiling hot water from aquifers which are profoundly covered in sedimentary basins of India. In the areas of normal or close typical temperature gradients, the most extreme temperature accessible in 3000 m wells is prone to be around 100 °C (Hutchence et al. 1986). Extensive research has been carried out on determining the parameters and processes affecting the geothermal extraction potential. Such research includes thermal, mechanical, and hydraulic (Te-Me-H) coupled numerical models as well as analytical studies.

Geological, geophysical, geochemical, geohydrological studies were carried out in the sites of geothermal reservoirs, but very limited studies were carried out to determine the potential of geothermal reservoir with doublet. From the limited studies carried out in India, it was observed that the maximum depth explored through drilling was about 362 m approximately which may not be sufficient to study the commercial feasibility of geothermal reservoir. This paper presents the simulation studies which are performed mainly to understand the movement of thermal front during the reservoir operation period which is also the cold water injection period. The study also presents the estimation of the probable production temperature that can be withdrawn from the production well under the cold water injection. The reservoir flow and energy transport model with doublet proposed in this study is suitable for a system which involves extraction of hot water from a geothermal production well with concurrent injection of cooled tailings water into the same formation in a nearby injection well. The sensitivity study of the energy production with respect to doublet distance and the cold water injection rate is also discussed in this study.

Case Study—Puga Geothermal Reservoir

Puga geothermal field, at an elevation of around 4400 m above mean ocean level, is situated in the northwest of the Himalayan ranges. Lying in the southeast part of Ladakh, the area of Jammu and Kashmir State, it frames a piece of the Himalayan geothermal belt, with topographical coordinates 33° 13′North and 78 °19′East as shown in the Fig. 1.

The Puga range is encompassed by slopes ascending to a height of around 6000 m, shaping a valley. This region is around 700 km far from Srinagar city and around 190 km from Leh Town, the district headquarters. Puga valley, situated in the northernmost territory, and is in the remotest and coldest part of the country and is around 15 km long with a most extreme width of 1 km inclining almost east west in bearing between Sumdo town in the east and Pologongka Lake in the west. The geothermal activity, which is spread over a range of 5 km^2 is confined to the two N–S drifting faults, known as Kaigar Tso flaw toward the west and Zildat short-coming toward the east. From the studies carried out by the geological survey of India, it was identified that the Puga geothermal region is the most prominent geothermal field in the nation, which has an expected capacity of 20–100 MW as found by the geo scientific studies for the zone.

The bore holes in the Puga valley is distributed at irregular intervals. The temperature gradient obtained in different well is different. However, due to the unavailability of proper temperature distribution for the entire valley, the decision maker finds it difficult to select the location of injection and extraction well. The

Fig. 1 Geographical location and the terrain of Puga Valley

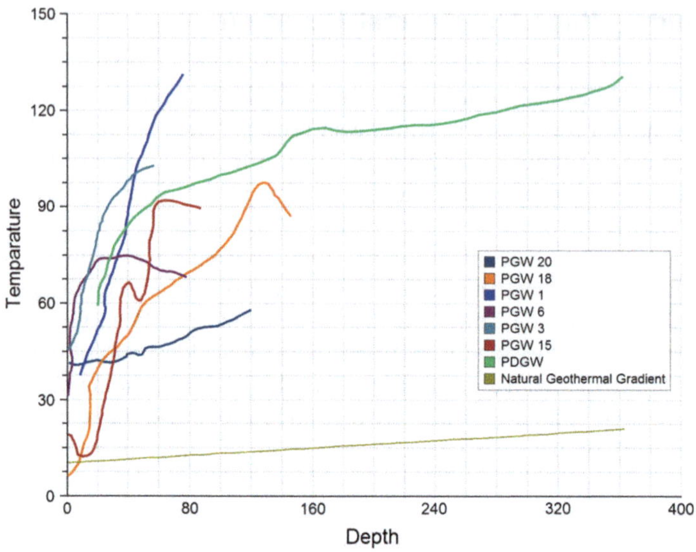

Fig. 2 Variation of temperature with depth at the corresponding bore holes

limited data of thermal gradient is available from the geothermal atlas of India (Geological Survey of India 1991) as shown in Fig. 2. For the present study the natural geothermal gradient (linear profile) as shown in Fig. 2, has been considered for the initial temperature distribution of the reservoir.

Conceptual Model of the Reservoir

The geophysical studies conducted in the Puga geothermal basin uncovered the presence of conductive features at shallow depths below 500 m, yet the deeper part of the geothermal valley remains unmapped. The wide band magneto-telluric studies were conducted in the Puga geothermal basin, to explore the data with respect to the vicinity of geothermal action and assets in that region (Harinarayana et al. 2006) and the computer model developed earlier by Absar et al. 1996. These studies were taken into consideration in the preparation of the present conceptual model of Puga geothermal valley. As the deep geothermal reservoir is emphasized in this study, the block heterogeneity is considered for the computer model to represent the Puga valley, approximately. To study the thermal front propagation under the injection and extraction wells, 3-D model of the valley is considered as shown in the Fig. 3.

The reservoir permeability is considered in terms of block heterogeneity based on the type of existing rocks in the Puga geothermal reservoir. From the literature (Absar et al. 1996) it is observed that the Puga geothermal valley is distributed with

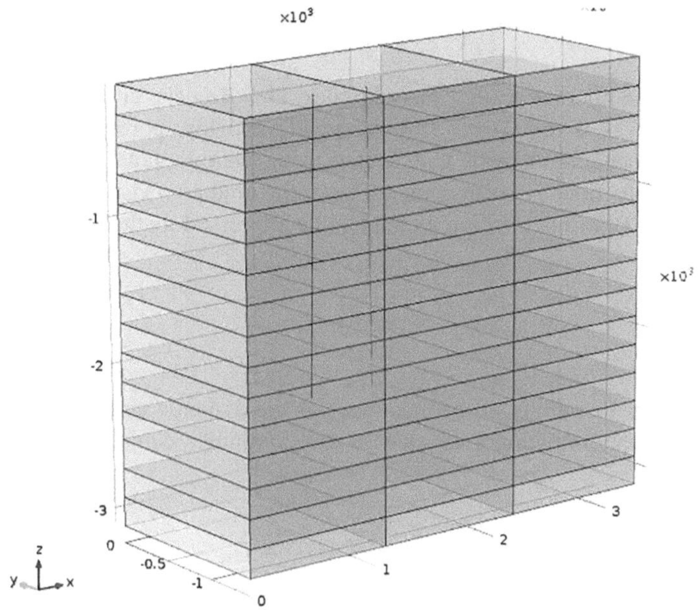

Fig. 3 Model approximation of entire Puga geothermal reservoir

Gneiss, Gneiss V, Gneiss III, Gneiss IV, Breccia, and Granite in the deep layers. Also the literatures suggest that the permeability of porous matrix follows log-normal distribution. Therefore, by considering the proper permeability range for the corresponding rocks, values are generated by following lognormal distribution which was further assigned for each of the blocks. The values of required properties for existing type of rocks are adopted from the literature (Robertson 1988; Manger 1963; Davis 1969; Sperl et al. 2008). Similarly, porosity, conductivity, density, and the specific heat capacity of each block is estimated and assigned to the respective block. Natural geothermal gradient is considered in determining the initial condition of the geothermal reservoir.

From Ahangar 2012, it is observed that the groundwater recharge to the Puga basin is mainly from the snow fed Puga–Nala and its tributaries. Puga valley has a stream coursing through it. This river streams from west to east and is nourished by ruling glacier, located at a distance of 15 km west of Puga. In the significant part of the area, with the exception of a small stretch in the western part, no point in the valley is more than 1000 m far from the stream. In the eastern part, the distances are much smaller. Flow in the Puga–Nala River is variable, i.e., 250 litres/s in the maximum peak summer and 4–6 L/s during December–January when the river is frozen.

Flow and Energy Transport Model with Doublet

As the Puga geothermal valley has alluvium and Breccia in its depth, the flow velocity is very less due to the less permeable and low porous material. In this study, time-dependent analysis is executed using Darcy's law. Darcy's law states that the velocity field is a function of pressure gradient, viscosity, and the soil properties of the porous medium which can be expressed mathematically as shown is Eq. 1.

$$u = \frac{K}{\mu} \nabla p \tag{1}$$

where k is the permeability of the medium, μ is the dynamic viscosity and ∇p is the pressure gradient. Equation 1 is used in the continuity equation given by Eq. 2 below.

$$\frac{\partial}{\partial t}(\rho \varepsilon_p) + \nabla.(\rho u) = Q_m \tag{2}$$

The side boundaries of the model are assigned with the Dirichlet boundary condition with specified hydraulic head. Bottom boundary is specified as no flow boundary. The model presented in this study considers the existing deep geothermal well of Puga valley as the injection well which is located at 740 m from left boundary of the model considered. And the extraction well is considered at a distance of 1033 m from the injection well.

Inlet condition to the injection well is assigned during the computation using the expression as shown in the Eq. 3.

$$-n \times \rho u = \rho U_0 \tag{3}$$

U_o represents the Darcy's velocity considered to be positive when the flow is inward and is considered to be negative when the flow is outward.

The transient heat transport model with conduction, convection, and heat source through porous media is taken into consideration to study the thermal front movement within the Puga geothermal basin as given by the Eq. 4.

$$d_z(\rho C_p)_{\text{effective}} \frac{\partial T}{\partial t} + d_z \rho C_p u.\nabla T = \nabla.(d_z k_{\text{eff}} \nabla T) + Q \tag{4}$$

where $(\rho)_{\text{effective}}$ is the effective density of the porous medium (fluid and solid), $(C_p)_{\text{effective}}$ is the heat capacity of the porous medium, k_{eff} is the effective thermal conductivity of porous medium, and Q is the heat source or a sink. The boundary

Table 1 Input parameters for the computer model

Parameter	Value	Units
Pumping rate	40	[L/s]
Radius of bore hole	0.5	[m]
Injection length	20	[m]
Production length	20	[m]
Injection velocity	0.0002	[m/s]
Production velocity	0.0002	[m/s]
Injection well inclination	1	[degree]
Extraction well inclination	1	[degree]
Hydraulic head gradient	0–5	[mm/m]
Surface temperature	290	[K]
Injection temperature	280	[K]
Mass influx	10	[Kg/s]
Heat source	60,000	[Watts]
Doublet distance	500	[m]

condition for heat transport model as given by Eq. 5 is geothermal gradient assigned to all the side boundaries of the model.

$$T = T_0, \text{ if } n \cdot u < 0; \quad -n \cdot q = 0, \text{ if } n \cdot u \geq 0 \tag{5}$$

The Darcy's flow model is coupled with the heat transport in porous medium model to study the temperature distribution under the operation of doublet.

Following the studies of (Blocher et al. 2010; O'Sullivan et al. 2001), this study is further extended to capture the behavior of geothermal reservoir as a 3-D model and the entire domain is divided into finer elements. The set of governing equations discussed above are solved numerically using COMSOL Multiphysics with the boundary conditions and input parameters discussed in the paper. To present the comprehensive model, all the parameters which had an impact on heat transport and fluid flow are also discussed. The concerned input parameters as well as geometrical parameters of the Puga geothermal reservoir are mentioned in the Table 1. The natural ground water flow is considered in terms of the specified hydraulic gradient as mentioned in the Table 1.

Results and Discussions

The governing flow and transport models are solved numerically using computational tool, COMSOL Multiphysics. As this study focus on the influence of geothermal doublet on the movement of thermal front, initially, the spacing

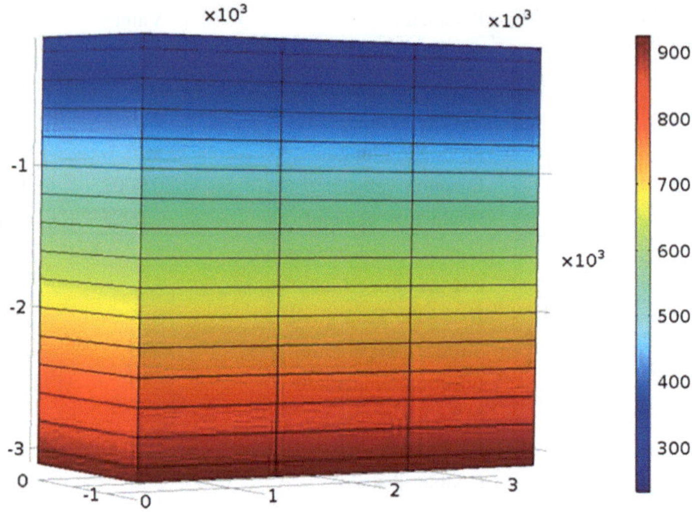

Fig. 4 Initial temperature distribution of the reservoir (natural geothermal gradient)

between injection and the extraction well is maintained at a distance of 500 m and then the thermal front variation is studied by varying the doublet distance to a lower value of 250 m and next to a higher value of 700 m, respectively. In addition to the well spacing or doublet distance, the impact of the injection rate of cold water is also studied by considering different injection rates of 50 and 30 L/s and compared with base simulation of 40 L/s.

It is observed that the scope of the most extreme and least temperature of the extracted water from the production well is increased with increase in the doublet distance. The initial temperature distribution is shown in Fig. 4. The movements of thermal front during the operation period for the different cases simulated are shown in Figs. 5 and 6. From the simulation studies, it is observed that the range of maximum temperature and the minimum temperature is increased with decrease in injection rate of cold water and also it is observed that with increase in the doublet distance, the time for which the cold water is in contact with porous structure which is initially at higher temperature increased which enhances the exchange of temperature for longer time, which apparently helps in extracting the water with high temperature at production well. The results of such study can help the decision maker in deciding the identification of proper locations of injection as well as the extraction wells in addition to the proper injection rates.

The pattern of thermal front along the vertical cross sections studied at different locations of the Puga geothermal region is shown in Fig. 7. The sections near to production well show higher temperature, as hotter fluid moves upward along the production well. Also the upward coning of the thermal front which is expected

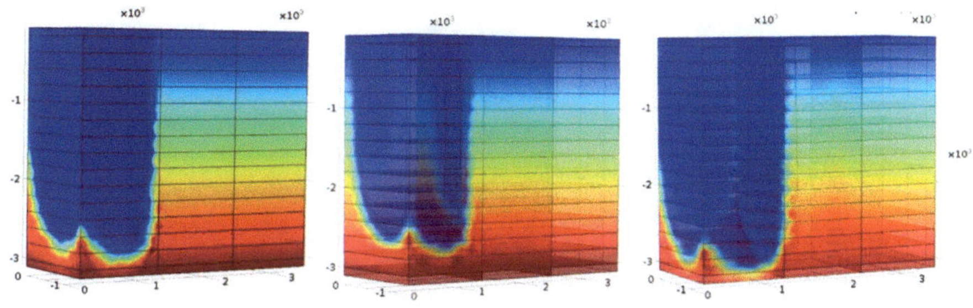

· **Fig. 5** Pattern of thermal front after 30 years and the impact of doublet distance (*left* 500 m, *center* 250 m, *right* 700 m)

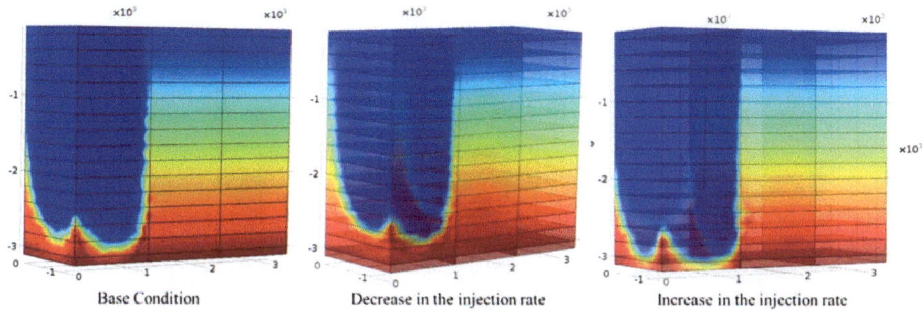

Fig. 6 Pattern of thermal front after 30 years and the impact of injection rate (*left* 40 L/s, *center* 30 L/s, *right* 50 L/s)

near the production well is shown by the results studied. The temperature of hot water at the production well over the operation period of 30 years in all the different cases is examined and is presented in Fig. 8. The temperature of extraction is higher in case of higher doublet distance and lower injection rate in comparison with the base condition considered in this study.

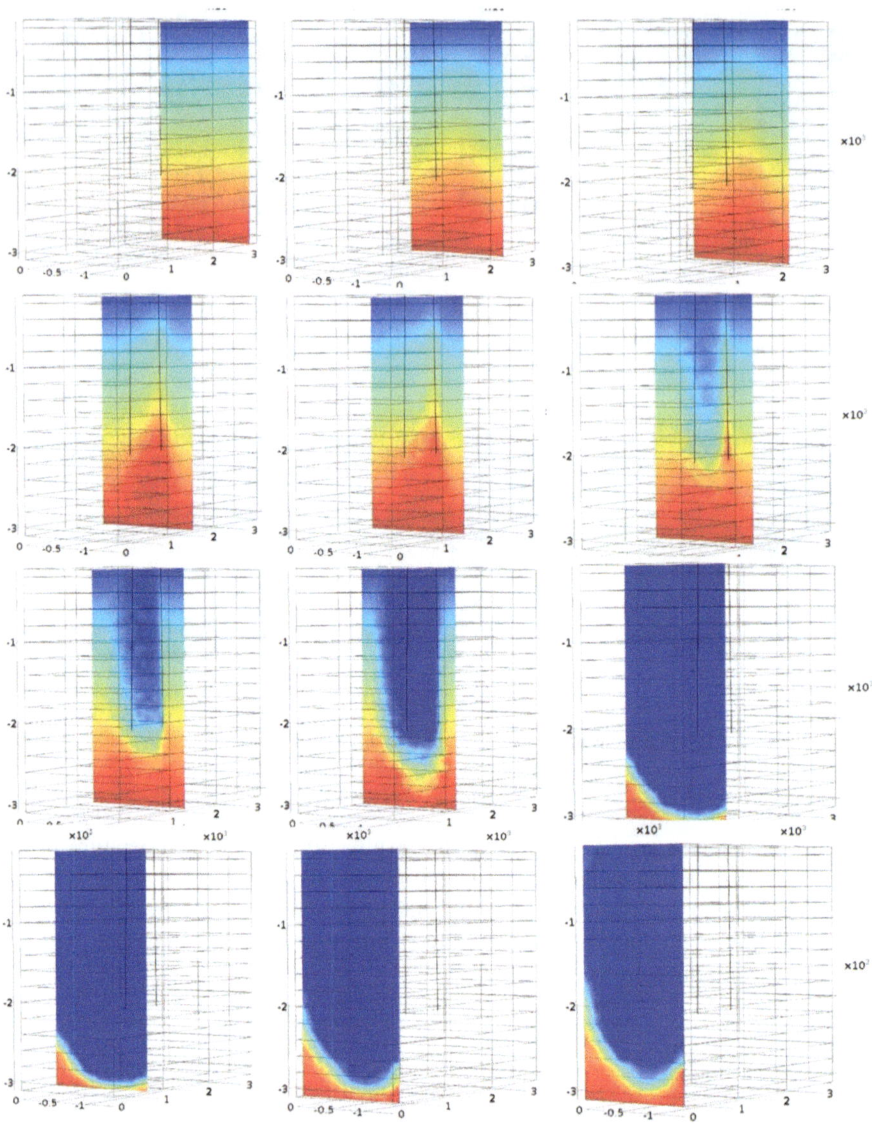

Fig. 7 Pattern of thermal front at different sections along the geothermal reservoir after 30 years

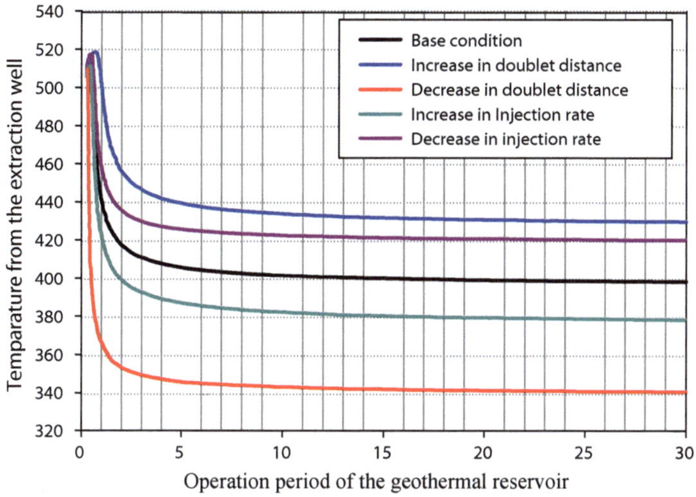

Fig. 8 Temperature of water that can be extracted through the production well

Conclusions

The primary focus of this study is to understand the thermal front movement under the doublet operation so that the potential of energy extraction in terms of temperature can be predicted for the geothermal reservoir of Puga region, Ladakh. This study helps in estimating temperature of water that can be withdrawn from the production well. The production temperature is required for further estimation of commercial energy generation purposes. It is found that the temperature extracted from the production well is proportional to the scale of electricity that can be generated. The behavior of Puga geothermal reservoir for the operation period 30 years is studied by considering some of the existing data for structural geometry and geological properties which is further studied by coupling the different physics of flow and energy transport involved during the operation period of geothermal reservoir. The results presented by such study will be helpful for the decision maker, to further plan and manage the injection and production rates, and the doublet distance, which has an impact on potential of geothermal reservoir.

References

Absar A, Kumar V, Bajpai IP, Sinha AK, Kapoor A, (1996) Reservoir modelling of Puga geothermal system, Ladakh, Jammu and Kashmir. Spec Publ Ser—Geol Surv India 45:69–74

Ahangar FA (2012) Feasibility study of developing a binary power plant in the low-temperature geothermal field in Puga, Jammu and Kashmir, India

Barbier Enrico (2002) Geothermal energy technology and current status: an overview. Renew Sustain Energy Rev 6(1):3–65

Blocher MG, Zimmermann G, Moeck I, Brandt W, Hassanzadegan A, Magri F (2010) 3D numerical modeling of hydrothermal processes during the lifetime of a deep geothermal reservoir. Geofluids 10(3):406–421

Davis SN (1969) Porosity and permeability of natural materials. Flow Porous Media 53–89

Ekneligoda TC, Min KB (2014) Determination of optimum parameters of doublet system in a horizontally fractured geothermal reservoir. Renew Energy 31(65):152–60

Geological Survey of India (1991) Geothermal atlas of India. Geological Survey of India

Harinarayana T, Abdul Azeez KK, Murthy DN, Veeraswamy K, Eknath Rao SP, Manoj C, Naganjaneyulu K (2006) Exploration of geothermal structure in Puga geothermal field, Ladakh Himalayas, India by magneto telluric studies. J Appl Geophys 58(4):280–295

Hutchence K, Weston JH, Law AG, Vigrass LW, Jones FW (1986) Modeling of a liquid phase geothermal doublet system at Regina, Saskatchewan, Canada. Water Resour Res 22(10): 1469–1479

Manger GE (1963) Porosity and bulk density of sedimentary rocks, No. 1144-E. *USGPO*

O'Sullivan Michael J, Pruess Karsten, Lippmann Marcelo J (2001) State of the art of geothermal reservoir simulation. Geothermics 30(4):395–429

Rao RUM (1997) Book review: geothermal energy in India (Geological Survey of India Special Publication, 45, 1996). J Geol Soc India 49:746–748

Robertson EC (1988) Thermal properties of rocks. No. 88–441. US Geological Survey

Schlagintweit R (1864) Thermal springs of India. J Asiatic Soc Bengal 33:49

Schulte T, Zimmermann G, Vuataz F, Portier S, Tischner T, Junker R, Jatho R, Huenges E (2010) Enhancing geothermal reservoirs. Geoth Energy Syst: Explor Dev Utilizat pp 173–243

Sperl J, Jirina T (2008) Permeability and porosity of rocks and their relationship based on laboratory testing. Acta Geodyn. Geomater 5(1):149

MIX
Papier aus verantwortungsvollen Quellen
Paper from responsible sources
FSC® C105338

If you have any concerns about our products,
you can contact us on
ProductSafety@springernature.com

In case Publisher is established outside the EU,
the EU authorized representative is:
Springer Nature Customer Service Center GmbH
Europaplatz 3, 69115 Heidelberg, Germany

Printed by Libri Plureos GmbH
in Hamburg, Germany